OXFORD STUDIES IN PHILOSOPHY OF SCIENCE

Science, Truth, and Democracy
Philip Kitcher

The Devil in the Details: Asymptotic Reasoning in Explanation, Reduction, and Emergence
Robert W. Batterman

Science and Partial Truth: A Unitary Approach to Models and Scientific Reasoning
Newton C. A. da Costa and Steven French

The Book of Evidence
Peter Achinstein

Inventing Temperature: Measurement and Scientific Progress
Hasok Chang

The Reign of Relativity: Philosophy in Physics 1915–1925
Thomas Ryckman

Inconsistency, Asymmetery, and Non-Locality: A Philosophical Investigation of Classical Electrodynamics
Mathias Frisch

Making Things Happen: A Theory of Causal Explanation
James Woodward

Mathematics and Scientific Representation
Christopher Pincock

Simulation and Similarity: Using Models to Understand the World
Michael Weisberg

Systemacity: The Nature of Science
Paul Hoyningen-Huene

Causation and its Basis in Fundamental Phyiscs
Douglas Kutach

Reconstructing Reality: Models, Mathematics, and Simulations
Margaret Morrison

The Ant Trap: Rebuilding the Foundations of the Social Sciences
Brian Epstein

Understanding Scientific Understanding
Henk de Regt

The Philosophy of Science: A Companion
Anouk Barberousse, Denis Bonnay, and Mikael Cozic

Calculated Surprises: A Philosophy of Computer Simulation
Johannes Lenhard

CALCULATED SURPRISES

CALCULATED SURPRISES

A Philosophy of Computer Simulation

Johannes Lenhard

OXFORD
UNIVERSITY PRESS

OXFORD
UNIVERSITY PRESS

Oxford University Press is a department of the University of Oxford. It furthers the University's objective of excellence in research, scholarship, and education by publishing worldwide. Oxford is a registered trade mark of Oxford University Press in the UK and certain other countries.

Published in the United States of America by Oxford University Press
198 Madison Avenue, New York, NY 10016, United States of America.

CIP data is on file at the Library of Congress
ISBN 978–0–19–087328–8

CONTENTS

PART II
CONCEPTUAL TRANSFORMATIONS

PART III
CONCLUSION AND OUTLOOK

ACKNOWLEDGMENTS

I am grateful to a number of close colleagues and friends who have given extremely helpful support at various stages during the present work. I wish to start by naming my academic teachers: Michael Otte, who guided me constructively in many ways as I moved from mathematics to philosophy; and Martin Carrier, who substantially encouraged and supported my move into the philosophy of science. His friendliness never impeded his critical attentiveness. I am grateful for major insights gained from discussions with Ann Johnson and Alfred Nordmann. When the manuscript was accepted by Oxford, I received three in-depth reviews that were of tremendous help and led to significant improvements to the text. Each review showed a remarkable amount of intellectual energy—a very agreeable experience for an author. My English tends to be clubfooted at times. Jonathan Harrow has done a supreme job in revising the text so that most traces of my German origins are concealed. I also wish to thank my illustrious colleagues at Bielefeld University's Department of Philosophy, Institute of Science and Technology Studies (IWT), Institute for Interdisciplinary Studies of Science (I^2SoS), and Center for Interdisciplinary Research (ZiF). Finally, it is a particular pleasure to thank my family, who so lovingly reminds me that life is so full of surprises—both calculated and uncalculated.

Major parts of some chapters have already been tested in lectures. To some extent, I have reorganized and supplemented material already used in published essays. Lenhard (2007) and Küppers and Lenhard (2005) have flowed into chapter 1; Lenhard (2009), into chapter 4; Lenhard and Otte (2010), into chapter 6; and Lenhard (2018), into chapter 7.

Substantial parts of most chapters are revisions of earlier German versions. They have been published in my monograph *Mit Allem Rechnen* (2015). I kindly acknowledge the permission of de Gruyter to modify and reuse this material.

CALCULATED SURPRISES

Introduction

What is calculable? This is not a new question—astronomic phenomena were being handled mathematically even in Antiquity. However, it became particularly relevant with the emergence of modern science, in which it was often related to the insight that motions on our planet follow mathematical regularities (Galileo) and that one can even formulate universal laws of nature that make no distinction between the physics of heaven and those of earth (Newton).

This approach of modeling something mathematically and thereby making it calculable was a major turning point in the emergence of our scientific world.[1] This is where computer simulation comes in, because it offers new recipes with which anything can be made calculable.

However, is computer simulation simply a continuation of existing forms of mathematical modeling with a new instrument—the digital computer? Or is it leading to a transformation of mathematical modeling? The present book addresses these questions. To extend the recipe metaphor, what is new about simulation has more to do with the recipe than with the actual ingredients. Nonetheless, anybody who is willing to try out a new recipe may well have to face a calculated surprise. Georg Christoph Lichtenberg (1742–1799) was already thinking along these lines when he wrote the following in one of his *Sudelbücher* (Waste books):

1. The classic accounts by Koyré (1968), Dijksterhuis (1961), or Burtt (1964) emphasize a standpoint viewing mathematization as the central feature of modern physics.

> How many ideas float about dispersed in my mind of which many pairs, if they were ever to meet, could produce the greatest discoveries. But they lie as far apart as the sulfur of Goslar from the saltpeter of the East Indies and the charcoal piles of the Eichsfeld, which when combined would produce gunpowder. How long must the ingredients of gunpowder have existed before gunpowder! (Lichtenberg, 2012, pp. 159–160, Waste book K 308)

Admittedly, there are also numerous recipe variants that are hardly worth discussing at all. Is computer simulation an excitingly new and powerful mixture or just a minor variation? A conclusive answer to this question requires a philosophical analysis that not only explores current forms of the scientific practice of computer modeling but also integrates the historical dimension of mathematical modeling. This is the only way to determine which continuities and differences are actually of philosophical interest. The key thesis in this book is that computer and simulation modeling really do form a new type of mathematical modeling.

This thesis consists of a twofold statement: first, that simulation reveals new, philosophically relevant aspects; and second, that these new aspects do not just stand unconnected alongside each other but also combine in ways that justify talking about simulation as a "type." Both elements of this premise are controversial topics in the philosophy of science. One side completely denies any philosophical newness; another side emphasizes single aspects as characteristics that—according to my thesis—reveal their actual type only when combined.

The widespread skepticism regarding any novelty is based largely on a misinterpretation. This views the computer as a logical-mathematical machine that seems only to strengthen trends toward mathematization and formalization. Viewing the amazing speed of computation as the main defining feature of the computer suggests such a premise as "everything as before, just faster and more of it." This ignores the philosophically relevant newness of simulation modeling right from the start. However, such a standpoint fails to recognize the transformative potential of computer modeling.

The far-reaching popularity of this standpoint is explained in part by the history of the use of the computer as a scientific instrument. Even when the computer reveals an unparalleled penetration rate and its use today has become a matter of course in (amazingly) many sciences, the situation was completely different in early decades.

In the pioneering days of the computer, a number of methods were developed that often laid the foundations for simulation procedures that have since come into common use. These include, for example, the Monte Carlo method (Metropolis & Ulam, 1949; von Neumann & Richtmyer, 1947) and the closely related Markov chain Monte Carlo method (Metropolis et al., 1953), cellular automata (Rosenblueth & Wiener, 1945; von Neumann, 1951), or artificial neural networks (McCulloch & Pitts, 1943). Other methods such as finite differences also received much attention and refinement through the application of computers. However, these contributions were of a more programmatic nature and were often ingenious suggestions for developing computer methods and techniques. Naturally, they were far in advance of contemporary scientific practice.

At that time, a computer was an exotic and enormous machine, a kind of calculating monster requiring much care and attention, and it was viewed with some skepticism by mathematically schooled natural scientists. Few fields had a need for extensive calculations, and these were also viewed as a "brute-force" alternative to a theoretically more advanced mathematization. Although the study of computer methods might not have been thought to reduce scientists to the rank of menial number crunchers, it did at least seem to distract them from more valuable mathematical modeling.

It took two further decades and the use of mainframe computers at computer centers throughout the world for a simulation practice to develop that would involve a significant number of researchers. Simulation now appeared on the scientific agenda, as witnessed by, for example, the annual Winter Simulation Conferences—international meetings of scientists from various disciplines who define themselves as both developers and users of simulation methods.

The need to consolidate the work of the growing group of computer modelers resulted in a few isolated works on computer simulation. These viewed the computer as being fully established as a new tool that would secure its own place in the canon of scientific methods. An early example is Robert Brennan's (1968) significant initial question in his title *Simulation Is Wha-a-t?* However, within less than one decade, Bernard Zeigler's book

Theory of Modelling and Simulation (1976) had already become a standard reference.

Starting roughly in the mid-1980s, computer-based modeling spread rapidly. The decisive factors underlying this development seem to have been the readily available computing capacity and the extended possibilities of interaction based on new visualization technologies. At the same time, semantically advanced languages were developed that opened up computer modeling to a large group of scientists who were not experts in machine languages. This merged both epistemological and technological aspects. With a rapidly growing store of available programs and program packets, science joined industry in making simulation and computer modeling a standard work approach. The topic of simulation was now visible from the "outside," as to be seen in the regular reporting on simulation-related topics in popular science magazines such as *Scientific American*.

One could expect this to have led to more analyses addressing the topic of simulation in the philosophy of science. However, that was not the case. I believe that this has been because the long-established, aforementioned view of the computer as a logical machine continued to suggest that its breadth and limitations had been determined for all time, so to say, by its logical architecture. My thesis, in contrast, is that mathematical modeling has undergone a fundamental change. To back up this diagnosis, I shall take examples from scientific practice and evaluate their methodological and epistemological implications. These examples assign a central role to modeling as an autonomous process that mediates between theory and intended applications. I shall discuss this further in this introduction.

First, however, I wish to return to the second part of my key thesis that simulation is not determined by one particular new feature or by the quantity of its epistemological and methodological features but, rather, by the ways in which these features are combined. Here, I can draw on the existing literature that has analyzed such characteristics of simulation.

Simulations only began to attract the attention of the philosophy of science from about 1990 onward. Paul Humphreys's and Fritz Rohrlich's contributions to the 1990 Philosophy of Science Association (PSA) conference can be seen as early indicators of this gradual growth of interest. For roughly two decades now, there has been a real growth in the literature on simulation in the philosophy of science. Modeling and simulation have become regular topics in specialist journals and they have their own series of conferences. There are now also publications on various aspects of simulation and computer modeling that address a broader audience from authors such as Casti (1997) or Mainzer (2003). However, the first monograph to take a philosophy of science approach to computer simulation came from Paul Humphreys (2004). This has been followed recently by the books of Winsberg (2010), Weisberg (2013), and Morrison (2014) that offer elucidating perspectives on the topic. A competently commentated overview of the most important contributions to the discussion in the philosophy of science can be found in Humphreys (2012).

An important starting point for the present work is the recent debate on models in the theory of science. This debate arose from the controversy over what role the theories and laws in the natural sciences take when it comes to applications.[2] Models, it has now been acknowledged generally for roughly a decade, play an important role by mediating between theories, laws, phenomena, data, and applications. Probably the most prominent presentation of this perspective can be found in Morgan and Morrison's edited book *Models as Mediators* (1999) that continues to be recognized as a standard work.

This model debate delivers an appropriate framework for simulation. I agree particularly with Margaret Morrison (1999) when she describes models as "autonomous agents." This role, I shall argue, is even taken to an extreme in simulation modeling. Their instrumental components along with their more comprehensive modeling steps make simulation models even less dependent on theory; they have become, so to say, more autonomous. The increasing differentiation of the model debate suggests that philosophical, historical, and sociological aspects could be considered together. The books edited by Gramelsberger (2011), Knuuttila et al.

2. Cartwright (1983) can be viewed as an important instigator of this controversy that I shall not be discussing further here.

(2006), Lenhard et al. (2006), and Sismondo and Gissis (1999) offer a broad spectrum of approaches to this debate.

The next question to ask is: Which particular features characterize simulations? The answers to be found in contemporary literature take different directions. The largest faction conceives the most important feature to be simulation experiments, often combined with viewing simulation as neither (empirical) experiment nor theory. Members of this faction include Axelrod (1997), Dowling (1999), Galison (1996), Gramelsberger (2010), or Humphreys (1994). Others consider the outstanding feature to be increasing complexity. Simulation and computer methods in general are treated as an adequate means of studying a complex world or, at least, complex phenomena within that world. Back in 1962, Herbert Simon already had the corresponding vision that has come to be viewed as standard practice since the 1990s. "Previously we had no models or frames or any set of mathematical equations for dealing with situations that combined two properties: a large number of individual parts, that is, degrees of freedom, and a complex structure regulating their interaction" (quoted in Braitenberg & Hosp, 1995, p. 7, translated). Closely related to this is the new syntax of mathematics—that is, the form of the models tailored to fit computers that moves away from the previously standard mathematical analysis. Instead, simulation draws strongly on visualizations (see, e.g., Rohrlich, 1991).

The emphasis on single characteristics is generally accompanied by a trend toward specialization. Most studies restrict their field of study in advance to a specific modeling technique. For example, Galison's (1996) claim that simulation represents a *tertium quid* between experiment and theory is based on a study of the early Monte Carlo method; Rohrlich (1991) or Keller (2003) see the true potential of simulation as being linked to cellular automata; whereas Humphreys (1991, 2004) or Winsberg (2003) basically derive their propositions from the method of finite differences. Others derive characteristics from the ways in which modeling techniques have become established in specific disciplines. For example, Lehtinen and Kuorikoski (2007) address agent-based models in economics, whereas Heymann and Kragh (2010) along with Gramelsberger and Feichter (2011) focus on the construction and use of climate models.

Without wishing to question the advantages of such a focused and particularistic approach, I wish to orient the present book toward more fundamental commonalities that hold for a large class of modeling techniques. Indeed, a thesis proposing a new type of mathematical modeling needs to possess a broad validity. Therefore, I place great value on basing my argumentation across the following chapters on a range of different modeling techniques and illustrating these with a series of examples from different scientific fields.

However, the task of characterizing simulation both epistemologically and methodologically raises a problem similar to that of having to steer one's way between Scylla and Charybdis: a too general perspective that views simulation as merging completely into mathematical modeling—and purports to see the computer as an instrument without any decisive function—is bound to be unproductive because the specifics become lost. Simulation would then seem to be nothing new—merely the mathematical modeling we have already known for centuries presented in a new garb. On the other hand, although a too detailed perspective would cast light on a host of innovations, these would remain disparate and end up presenting simulation as a compendium of novelties lacking not only coherence but also any deeper relevance for the philosophy of science. Accordingly, one challenge for the present study is to select both an adequate range and an adequate depth of field.

Hence, my starting point is determined in two ways: First, I focus attention on the *process* of modeling, on the dynamics of construction and modification. Second, I take *mathematical* modeling as a counterfoil. Here, I am indebted mostly to the work of Humphreys (1991, 2004) that convinced me of the value of this counterfoil approach. My main thesis is that by taking the form of simulation modeling, mathematical modeling is transformed into a new mode. Or to put it another way, simulation modeling is a new type of mathematical modeling. On the one hand, this means that simulation is counted into the established classical and modern class of mathematical modeling. On the other hand, a more precise analysis reveals how the major inherent properties of simulation modeling contribute to a novel explorative and iterative mode of modeling characterized by the ways in which simulation models are constructed and fitted.

Hence, as stated, I shall try to steer my way between Scylla and Charybdis—that is, to avoid taking a perspective that is too general because it would then be impossible to appreciate the characteristics of simulation, but to simultaneously avoid selecting a focus that is too specific because this would make it necessary to address each different technique such as Monte Carlo or cellular automata separately. Admittedly, this will inevitably overlook several interesting special topics such as the role of stochasticity (implemented by a deterministic machine)—but even Odysseus had to sacrifice six of his crew.

What is it that makes simulation modeling a new type of modeling? First, it has to be seen that this question addresses a combination of philosophy of science and technology. Not only can mathematical modeling address the structure of some phenomena as an essential feature of the dynamics in question; it can also apply quantitative procedures to gain access to prediction and control. This general description then has to be modified specifically owing to the fact that the modeling in which I am interested here is performed in close interaction with computer technology. One direction seems self-evident: the (further) development of computers is based primarily on mathematical models. However, the other direction is at least just as important: the computer as an instrument channels mathematical modeling.

Initially, this channeling can be confirmed as an epistemological shift. Traditionally, mathematical modeling has been characterized by its focus on the human subjects who are doing the active modeling in order to gain insight, control, or whatever. Now, an additional technological level has entered into this relationship. Mathematical modeling is now (also) performed with computers; it integrates computers into its activities: the adequacy of models is measured in terms of the phenomena, the intended application, and then the instrument as well. However, this subjects mathematization to a profound change. Throughout this book, it will often become apparent how a dissonance emerges between continuity with the older type of mathematical modeling and the formation of the new type. Simulation has to achieve a sort of balancing act that compensates for the transformations introduced by the computer, making it, so to speak, compatible through additional constructions within the model.

A further aspect of this channeling that needs to be considered here is the special relationship to *complexity*. In the twentieth century, this term grew into a central concept in the sciences and it has been shaped decisively through computer-based modeling. Complex systems are considered to lack transparency, because their components define the system dynamics only after a multitude of interactions. This somewhat vague paraphrase is rendered more precise by the specific and defined meaning of *computational complexity* (proceeding from the work of the mathematician Kolmogorov). This complexity sets a limit on computer models in the sense that complex problems either cannot be solved or—depending on the degree of complexity—can be solved only slowly by computational power. Computers consequently stand for an unfettered increase in computational power, but nonetheless are restricted to the domain of simple or less complex problems. The groundbreaking insights of the meteorologist Lorenz have led to recognition of how relevant complex systems are in natural events. Parallel to this—and that is what is important—computer models themselves are complex: complexity characterizes both the object *and* the instrument.

Broadly speaking, complexity is not restricted to exceptionally large systems that are studied with computer methods. The degree and extent of complexity are far more comprehensive and have permeated into broad areas of simulation modeling itself.

Therefore, studying the dynamics of the system being examined and also those of the model itself become separate tasks. The specific reflexivity of simulation is documented in the way that this type of mathematical modeling has to mediate between an excess of power measured in "flops" and a multifaceted problem of complexity that limits the range of this power.

In many cases, it is not possible to reduce complexity: This would be neither appropriate for the phenomena concerned nor methodologically performable. This is what changes the core activity of mathematical modeling. It has less to do with overcoming complexity than with handling it. In other words, simulation modeling is a mode of modeling that balances high complexity with usability—even when this makes it necessary to abandon some of the "classical" virtues of mathematical modeling.

The main characteristics of simulation modeling presented as follows depend on and mutually reinforce each other. This delivers the stability that enables them to build a new type of modeling. Each of the first four chapters of part I deals with one of the following elements of simulation modeling.

- *Experiment and artificiality*: This is where the channeling effect of the technological level is revealed, because computers can process only discrete (stepwise and noncontinuous) models. In contrast to traditional mathematical modeling, such models seem artificial. If a continuous theoretical model is available that is then redesigned as a computer model, this discretization results in a flood of local interactions (e.g., between finite elements). The model dynamics then have to be observed in a computer experiment. Vice versa, in the course of the modeling process, the simulated dynamics can be approximated to theoretically determined dynamics or to known phenomena using special built-in parameters. This requires the addition of artificial components—that is, components that are formed for instrumental reasons.
- *Visualization and interaction*: Exploiting the enormous processing capacity of computers generally requires a powerful interface on which it is possible to prepare very large datasets for researchers in ways that enable them to estimate what they will then need for further stages of construction. Visualization offers an excellent possibility here, because of the quite amazing human capacities in this field. When the modeling process calls for the repeated running of an adjustment loop, this interface is correspondingly of great importance.
- *Plasticity*: This describes a property of models. Some classes of models intentionally use a weak structure in order to exploit not their fit but their ability to fit. One can then talk about structural underdetermination. In other model classes, the dynamics are codetermined strongly by a theoretical structure. Even when such a model correspondingly cannot be taken as being

structurally underdetermined, plasticity can still be an important feature. It then draws on the effects that arise from the ways in which the (artificial) parameters are set. The more flexible a model is, the more significant is the phase of modeling during which the parameters are adjusted. Put briefly, the plasticity of a model is proportional to the breadth of an iterative and exploratory adjustment during the modeling.

- *Epistemic Opacity*: This kind of opacity has several sources. By their very definition, complex models are nontransparent. Things are similar for case distinctions that are very simple for a computer to process algorithmically, but practically impossible for human beings to follow. Moreover, the artificial elements discussed in chapter 1 contribute to opacity, and these are added particularly for instrumental reasons and in response to the performance observed—that is, for nontheoretically motivated reasons. The value of such elements first becomes apparent during the further course of the modeling process. This lack of insight into the model dynamics is a fundamental difference compared to traditional mathematization that promises to produce transparency through idealization. Things are channeled differently in simulation modeling, and the iterative sounding out of the model dynamics is applied as a methodological surrogate that should replace intelligibility.

Chapter 5 summarizes how these components all fit together to form an iterative and explorative mode of modeling. Even when the proportions of these components may vary from case to case, it is nonetheless their *combination* that confirms simulation modeling to be an independent and novel type. Simulation models develop their own substantial dynamics that are derived not (just) from the theories entered into them, but (also) are essentially a product of their construction conditions. The emphasis on these intrinsic dynamics and the way the construction mediates between empirical data and theoretical concept place computer simulation in the framework of a Kantian epistemology.

I agree with Humphreys (2009) when he refers to the concept of the "style of reasoning" when categorizing simulation.[3] Based on the results discussed in this book, simulation modeling has to be specified as a *combinatory style* of reasoning. The new mode of modeling is having—I wish to argue further—a major influence on the progress of science; namely, it is pointing to a convergence of the natural sciences and engineering. This contains both instrumental and pragmatic approaches focusing on predictions and interventions. Simulation modeling brings the natural sciences and engineering more closely together into a systematic proximity.

Part II is dedicated to the conceptual shifts triggered by the establishment of the new mode or style. The extent of the transformative power that simulation and computer modeling are exerting on traditional mathematical modeling can be pursued further on the basis of two concepts. Chapter 6 discusses the concept of solution that is actually formulated very clearly with respect to an equation or, at least, to a mathematically formulated problem. In the course of simulation, the talk is about a numerical solution that is actually not a solution at all in the older, stricter sense but, rather, something that has to be conceived in a highly pragmatic way. The concept of the solution is then no longer defined according to criteria coming purely from within mathematics, but instead according to whether the numerical "solution" functions adequately enough relative to some practical purpose. I shall draw on the dispute between Norbert Wiener and John von Neumann when discussing the dissonance between a standpoint that views numerical solutions as being derived from mathematical ones and the opposite position that regards numerical solutions as existing in their own right.

When it comes to applications, one central problem is how to validate a model. Simulation represents a kind of extreme case regarding both the number of modeling steps and the complexity of a model's

3. The differentiation of scientific thinking styles goes back to Alistair Crombie (1988; see also, for more detail, 1994). Ian Hacking specified the concept more rigorously as "style of reasoning" (1992). The main point is that such a style first specifies "what it is to reason rightly" (Hacking, 1992, p. 10). "Hypothetical modeling" is one of the six styles identified by Crombie.

dynamics. Does this create new validation problems? I address this issue in chapter 7, in which I present a two-layered answer. On the first layer, the validation problem remains the same: the extent of the modeling simply makes it more precarious. On the second layer, a conceptual problem emerges: simulation reaches the limits of analysis in complex cases, and this leads to the problem of confirmation holism. A key concept here is modularity. Organizing complex tasks into modules is seen as the fundamental strategy when working with complex designs. I argue that the modularity of simulation models erodes, for systematic reasons. As modeling proceeds, submodels in the planning phase that were clearly circumscribed almost inevitably start to overlap more and more. The practices of parameterization and kluging are analyzed as principal drivers of this erosion. These limit the depth of the analysis of model behavior; that is, under some circumstances, it may no longer be possible to trace individual features of behavior back to individual modeling assumptions. The only method of validation this leaves is a (stricter) test on the level of the model's behavior—as performed in technological testing procedures.

Part III, which is the final chapter 8, addresses two issues. First, it draws a conclusion from the autonomy of simulation modeling and what this means in terms of novelty. Based on my previous analyses, I wish to defend a moderate position. The newness of simulation modeling consists essentially in the ways in which it constructs mathematical models and operates with them. This is the source of its potential as a "combinatorial style of reasoning." At the same time, I reject the exaggerated demands and claims often linked to the topic of simulation—for example, that modern science refers only to a world *in silico* nowadays, or that simulation has finally given us a tool with which to unveil complexity. Simulation modeling functions in an explorative and iterative mode that includes repeated revision of the results. Or, to put it another way, what is certain about simulation is its provisional nature.

Second, chapter 8 offers an outlook for a philosophical and historical thesis on the relation between science and technology. Chapter 5 already related the mode of simulation modeling to the convergence of the natural sciences and engineering. It is particularly the types of mathematization striven toward in both domains that are drawing closer together. This movement affects the concept of scientific rationality. I illustrate this claim by looking at the divide between a realist and an instrumentalist standpoint. Does science flourish by subscribing to theoretical insight or, rather, to material manipulation? Often, this divide is treated as irreconcilable. Simulation modeling, however, changes the picture because it intertwines theory, mathematics, and technological instrumentation, thus making the divide look much less severe and significant. Finally, the analysis of simulation modeling suggests the need to rethink what constitutes modern scientific rationality.

PART I

A NEW TYPE OF MATHEMATICAL MODELING

1

Experiment and Artificiality

SIMULATION AND EXPERIMENT

Computer simulations work with a new type of experiment—on that, at least, the literature agrees. Indeed, it is precisely this new type of experiment that frequently attracts philosophical attention to simulations (Dowling, 1999; Gramelsberger, 2010; Humphreys, 1991, 1994; Hughes, 1999; Keller, 2003; Morgan, 2003; Rohrlich, 1991; Winsberg, 2003) and how they compare with thought experiments (see Di Paolo, Noble, & Bullock, 2000; or Lenhard, 2017a). Although there is consensus that one is dealing with a special concept of the experiment here insofar as it does not involve any intervention in the material world, such agreement does not extend to what constitutes the core of such an experiment. As a result, no standardized terminology has become established, and simulation experiments, computer experiments (or, more precisely, *computational experiments*), numerical experiments, or even theoretical model experiments all emphasize different, though related, aspects. The aim of this chapter is not to elaborate a critical taxonomy. Instead, it concentrates on those aspects of experimentation that are essential to the new type of modeling.

It may well seem attractive to "experiment with ideas"—and particularly to do this in a controlled and reproducible way. However, simulation experiments are not carried out in the realm of ideas but, rather, on a computer. This constitutes an essential condition, because such experiments always require a numerical form. The iterative processing of discrete units is quite simply one of the conditions making it possible to use the computer as a tool. Even when such formal conditions may seem unspectacular, they have a major effect. One could compare this with the algebraic thinking introduced by Descartes, Fermat, and others. Although algebra started as a computing technique, it soon developed the revolutionary effects on mathematics, as well as on the broader realm of philosophy portrayed so convincingly by Michael Mahoney (1991). In this first chapter, my aim is to show how computer experiments are now contributing to a further new transformation.

It would be misleading to view such experiments as the sole feature defining simulations. This would inexcusably neglect a host of other important features and make their characterization unidimensional. This chapter will relate the experimental part of simulation modeling to a specific artificiality of mathematical modeling that stems from the discrete (noncontinuous) character of computer models.

In his classic *Representing and Intervening*, Ian Hacking (1983) juxtaposed two major domains of scientific methodology that he associated with either theoretical construction or experimental access. From such a perspective, simulation would seem to be some sort of hybrid. Although a material instrument, the computer simultaneously executes symbolically compiled programs and can function as a tool with which to implement and analyze theoretical models. Simulation experiments are evidently performed in both of Hacking's domains, and this alone shows how they already represent a conceptual shift in the idea of the experiment. However, such simulations are not interventions in nature but, rather, experimentations with theoretical models, equations, or algorithms that also occur within or with the computer. When the talk is about computer experiments, the object of attention is the behavior of theoretical and symbolic entities as seen "in" the computer—that is, within a concrete implementation on a material electronic machine.

A purist might well object that the term "experiment" should be reserved for situations that investigate nature (even if in an artificial environment). On the one hand, it is certainly true that there are some contexts in which all that is meant is an experiment in this sense—for example, in Hacking's (1983, chap. 16) instrumental argument for realism.

On the other hand, it has to be acknowledged that numerical experiments function like experiments in at least one important aspect—namely, when researchers are observing how simulation models behave under varying initial and framing conditions, as in an experimental design. I consider it to be completely legitimate to talk about a type of experiment here, especially in light of all the other established ways of talking about an experiment, such as talking about a thought experiment, that also ignore this puristic restriction. The books edited by Gooding et al. (1989), Heidelberger and Steinle (1998), or Radder (2003) provide a good impression of the great breadth of conceptions of the experiment, and the last-mentioned book even contains two articles on simulation experiments.

Even in the early days of computers, it soon became established to talk about experiments as one possible field in which to apply them. For example, right from the start, the mathematician Stanislaw Ulam, one of the pioneers of computer methods, viewed the invention of the Monte Carlo simulation (to which he had contributed decisively) as a new method for carrying out statistical and mathematical experiments (Ulam, 1952; Metropolis & Ulam, 1949). Trials were carried out with iterated computer runs that each began with (pseudo)randomly varying parameters. A computer can be used to perform a great number of such single runs. Then, in line with the law of large numbers, one can view the mean of the results as an approximation of the expected value (in a probabilistic sense). The advantage is that it is easy to compute the mean from the single results even when the expected value itself is analytically inaccessible. As a result, one can use computational power to perform mathematical analyses as long as it is possible to present the value concerned as an expected value. Linking this to the stochastic theory of Markov chains led to its further development as the Markov chain Monte Carlo method (already established by Metropolis et al. in 1953). This can be used to approximate a surprisingly large class of integrals numerically, thereby granting this experimentally supported method a very wide range of applications.

However, lurking behind such experiments one can find a profound set of problems in the philosophy of mathematics, because although the (pseudo)random numbers generated by the computer are numbers that share the same characteristics as random numbers, they are generated by a deterministic machine (see, e.g., Kac & Ulam, 1968). Moreover, the last few years have seen the emergence of an "experimental mathematics" that aims to complement theoretical penetration through experimental testing by computers. This could turn into an autonomous subdiscipline that would undermine the self-conception of mathematics as an a priori science (Baker, 2008; Borwein & Bailey, 2004; Borwein, Bailey, & Girgensohn, 2004; van Kerkhove & van Bendegem, 2008).

Have I now covered all the major points when it comes to characterizing simulation? Peter Galison (1996) sets out such a premise in his contribution to a highly informative and readable study on the historical origin of the Monte Carlo simulation within the context of the Manhattan Project. He assigns a fundamental philosophical significance to simulation in this context because of the way it is linked to a radically stochastic worldview. Although this may hold for some areas of Monte Carlo simulation, Galison's premise depends crucially on its limitation to this one method. Other equally important and, indeed, even more widespread simulation methods also work with experiments without, however, assigning any special role to chance. Therefore, despite the fact that "statistical" experiments represent an important and philosophically interesting class of computer experiments, any analysis of the role of simulation experiments requires a broader focus.

In the following, I am not interested in localizing simulation experiments as a kind of experiment. There is already sufficient literature on this, as mentioned earlier. My goal is far more to work out in what way or ways experimenting is fitted into the process of simulation modeling: How far do numerical experiments contribute to making simulation modeling a special type of mathematical modeling? My main point is that the discreteness of the computer makes it necessary to perform repeated experimental adjustments throughout the modeling process. Experimental practice (in the ordinary sense) is bound up with adjustments such as calibrating instruments. With simulation, they become essential to mathematical modeling, as well.

THEORETICAL CHARACTER

Speaking about theoretical model experiments expresses an aspect that I consider to be of central significance. It is simulation *models* that are being experimented with. Certainly, these are not always finished models that are intended to deliver results. Typically, they are models at a preliminary stage whose further construction is based essentially on experimental analyses. Therefore, the experimental character is found not just during the runs of the final simulation model but also throughout the modeling process.

My emphasis on the activity of modeling and the process of model construction is completely intentional. This is because it is major points within this process that take a different form in simulation modeling compared to the usual mathematical modeling—not least because, as I shall argue, taking experimental paths is not only possible but also essential. The usual presentations of simulation modeling neglect this aspect and separate the formulation of the model from its experimental testing. This makes it seem as if the only analysis that is being performed experimentally in the simulation is the final one. From the host of such accounts, one could single out Zeigler (1976) as a standard reference. On the philosophical side, this view corresponds to, for example, Hughes's (1997) DDI (denotation, demonstration, interpretation) account of modeling in which, first, something is represented in a model; second, conclusions are drawn in the model; and third, these conclusions are transferred to the phenomena being modeled.[1] In such a view, simulation models in particular do not differ in their formulation from (mathematical) models in general. Hence, a more precise analysis is required if the specifics of simulation modeling are to emerge.

Theoretical and experimental approaches are combined or, perhaps more aptly, interwoven with each other. This leads the autonomous and mediating role of simulation modeling to emerge particularly clearly. I shall address the strong relations to Margret Morrison's (1999) view

1. Similar positions can be found in, for example, Neelamkavil (1987) or Stöckler (2000).

of "models as autonomous mediators" toward the end of this chapter. She used Prandtl's fluid flow model to support her standpoint that this had animated the introduction of the boundary layer through which the theory of fluid mechanics gained a wide range of applications (see also Heidelberger, 2006, for this example).

Theoretically, fluid mechanics is a well-researched area; at the same time, however, it is also a prominent case in which analytical solutions often are restricted to a local nature or might not be attainable at all. Although the Navier–Stokes equations are formulated with the tools of differential calculus, they remain (analytically) unsolvable with these tools. Here, Prandtl's model was able to mediate among the theory, the data, and the phenomena—and thus become a prime example of a mediating model.

The relevance of fluid mechanics for applications has now undergone a further enormous boom through simulation methods. And precisely because it is based so strongly on theory, this presents an excellent case for a philosophical discussion of simulation. This is because that which can be confirmed here about the role of experimenting and testing applies a fortiori to fields that are based less strongly on theory. In those fields in which simulations work with a completely new syntax, such as that of cellular automata, there is in any case a lack of alternative powerful mathematical tools that can be used to find out something about how the model behaves. In this situation, there is basically no alternative to the simulation experiment. In theoretically based cases such as fluid mechanics, however, this is not so evident. Then it is easy to view simulation models as dependent on theory and as merely technical computing aids—and anything but "autonomous."

Many fields in the natural sciences and particularly in physics use differential calculus to gain a mathematical description of the phenomena observed and to formulate the underlying laws. Indeed, long before the computer age, it was understood that complex phenomena described with systems of interacting equations are almost impossible to handle analytically. Scientists were compelled to give a wide berth to such "unsolvable" systems.

Typical examples of this class are phenomena in fluid mechanics, and I would like to take the global circulation of the atmosphere as my example here and use it here for a detailed case study.

Atmospheric dynamics encompasses both the weather and the climate. Whereas the weather forms what is called a *chaotic system* (which I shall be discussing in the next chapter) that renders long-term predictions impossible, the climate is something like "average" weather. It displays several well-known and stable phenomena such as the wind zones that develop to the north and to the south of the equator.

Initially, this distinction was anything but evident. Vilhelm Bjerknes, a student of Heinrich Hertz, already postulated a theoretical model for the circulation of the atmosphere that should make it possible to predict the weather on the basis of the known laws of motion and fluid mechanics (Bjerknes, 1904). This remained a model. Although Bjerknes was able to show that the number of equations and unknowns could "merge," he was unable to derive any predictions. Lewis F. Richardson (1922) performed further pioneering work. He revisited Bjerknes's model. Formally, it was a system of partial differential equations that should describe (at least in theory) the known phenomena of the atmosphere. However, the system is "analytically unsolvable"—that is, one cannot obtain a solution by integrating it with the traditional methods of mathematics. Such a solution would describe the behavior ensuing from the interdependency of the equations (interaction of laws).

Richardson's groundbreaking innovation was to bypass the analytical inaccessibility by applying numerical methods. He replaced the continuous dynamic with a discrete, stepwise version in which the differential equations were transformed into difference equations on a grid. The resulting system of difference equations in which, roughly speaking, infinitesimal intervals were replaced by finite ones and integration became summation, should, so was his idea, approximate the original system while simultaneously making it possible to master the complex dynamics through continuous iteration of the equations (instead of integration). This required a great deal of calculation, although it consisted mainly in an enormous number of very elementary operations. One can almost see an industrial solution program here. Richardson imagined a veritable "factory" with many employees.

Indeed, he anticipated that more than 60,000 people would be needed to work out the complete dynamics of the theoretical model through calculation by hand (including mechanical desktop machines, but before the invention of the electronic computer).[2]

Although Richardson is considered to be the father of numerically based weather prediction, this is basically in recognition of his plans rather than his successes. In fact, he produced only one single prediction that took him months to compute—and turned out to be utterly wrong. His predictions were of no use. As it turned out later, this was not just due to a lack of computational power. His discretization procedure was already problematic, because not every oversimplification through a grid will produce an emerging dynamic that is qualitatively close to the initial continuous dynamic. However, such considerations emerged only after computers became available. And, I should like to add, it was only then that they could emerge, because this would require experimental opportunities that were not yet available to Richardson, who had to develop and run his discrete model completely by hand.

Modeling would first function only when the modeling process could be linked closely to experiments. This refutes a common and influential belief about simulation models based on the incorrect idea that the discrete versions of models depend completely on the theoretical models. Purportedly, all that is required is one new intermediate step that will translate continuous models into the discrete world of the computer. Accordingly, the discrete model was—and repeatedly continues to be—understood as a mere schematic aid applied to numerically compute the solution to the theoretical model.

In the following, I shall further develop this example and show how a change of perspective was required. It was necessary to move away from simulation as a mere numerical annex to the equations in a theoretical model, and to start to view it as an independent type of mathematical modeling that, so to speak, takes advantage of the specific strengths of this instrument.

2. The word *computer* came into general use during World War II, when it was used initially to describe a mostly female person who performed such activities. The military ran several computing factories in the United States. See David Grier's (2005) study, *When Computers Were Human.*

A FIRST EXPERIMENT

Nowadays, simulation models are accepted as a central part of weather prediction and climatology. What makes the use of computers so successful? And what role does experimenting play in this? I shall concentrate completely on the global dynamics of the atmosphere. During the decades before the development of the computer, there had already been some successes in modeling atmospheric dynamics (see the account in Weart, Jacob, & Cook, 2003). However, these addressed only some aspects of global circulation such as *lateral diffusion* (Rossby in the 1930s) or the *jet stream* (Palmèn and Riehl in the 1940s). The majority of meteorologists were, in any case, skeptical about any possibility of modeling the complete system. The predominant opinion was "that a consistent theory of the general circulation is out of reach" (Lewis, 1998, p. 42, referring to Brunt, 1944). If that were to have been the case, the path taken by Bjerknes and Richardson would have been a dead end. It took the availability of a new instrument— namely, computer simulation—to change this skeptical appraisal.

One of the main contributors to the conceptual development of the computer was the Hungarian mathematician John von Neumann.[3] This immigrant from Germany to the United States was also a driving force in framing applications for this new mathematical machine during World War II and the period immediately after. "To von Neumann, meteorology was par excellence the applied branch of mathematics and physics that stood the most to gain from high-speed computation" (Jule Charney, quoted in Arakawa, 2000, p. 5). With von Neumann as mentor, a meteorological research team was assembled at the Institute for Advanced Studies (IAS) in Princeton headed by Jule Charney, and this formed the nucleus for the Geophysical Fluid Dynamics Laboratory (GFDL) founded in Princeton in 1960—the first institute focusing on simulation-based meteorology or, as it was later called, climatology. The declared goal was to model the

3. Akera's monograph (2006) delivers an excellent insight into the complex theoretical and technical history of this development of the computer involving so many different persons and institutions. William Aspray's (1990) book is still the standard work on the significance of von Neumann's contribution.

fluid processes in the atmosphere and to "solve" the corresponding differential equations on the computer; in other words, basically to use the new instrument to transform Richardson's failure into a success. The name "Geophysical Fluid Dynamics Laboratory" already conveyed the orientation toward fluid mechanics and simultaneously the self-definition as an experimental computer laboratory.

The Charney group conducted simulations of atmospheric conditions. Norman Phillips worked in this group at the IAS, and he was the one who transferred Richardson's primitive equations into a computer model with an eye toward producing global circulation patterns. In 1955, he actually managed to simulate the dynamics of the atmosphere in the so-called *first experiment*—that is, he successfully reproduced the wind and pressure relations of the entire atmosphere in a computer model.[4] Naturally, this required him to reformulate the continuous equations of motion and hydrodynamics in a way that would enable him to compute them at the grid nodes. The group at IAS followed a more appropriate strategy than that pursued by Richardson. In particular, they were now aware of the Courant–Friedrichs–Lewy condition with which the discretization has to comply in order to prevent the discrete dynamics from running out of control. However, whether and how far the conditions were sufficient remained an open issue.

This modeling step represents an unavoidable conceptual task because of the discrete logic of the new instrument—the computer or computer simulation. The next step in the simulation experiment was to add the dynamic forcings—that is, the radiation of the sun and the rotation of the earth. Driven by these conditions, the atmosphere left its state of rest and settled into a *steady state* corresponding to a stable flow pattern—or, to put it more precisely, the simulation created the impression of doing this. Indeed, even the question when one can talk about such a stationary state and view it as more than an intermediate stage in the simulation model is one that can be resolved only by analyzing the generated (simulated) data. In fact, Phillips's estimates were way off. As a follow-up study showed, his model took far more time to settle than Phillips had assumed

4. See, for a detailed description, Lewis (1998); see, for a more far-reaching history of the ideas behind the modeling of the general circulation of the atmosphere, Lorenz (1967).

(Wiin-Nielsen, 1991). The decisive question was whether the model would adequately reproduce the global pattern of wind streams observed in the real atmosphere. This meant comparing the model pattern with the pattern and flows in the atmosphere known from the long history of observing the weather, particularly in seafaring. These included the west winds north of the equator—the *surface westerlies*—whose pattern is illustrated in figure 1.1.

Over the course of the simulation, the atmosphere left its initial state of rest and moved into the "right" type of circulation pattern (Ferrel type):

> the zonally averaged meridional circulation in the middle latitudes changed from the Hadley type to the Ferrel type, producing the midlatitude surface westerlies. In this way, the experiment simulated the very basic features of the observed general circulation of the atmosphere, whose causes had been more or less a matter of speculation. (Arakawa, 2000, p. 8)

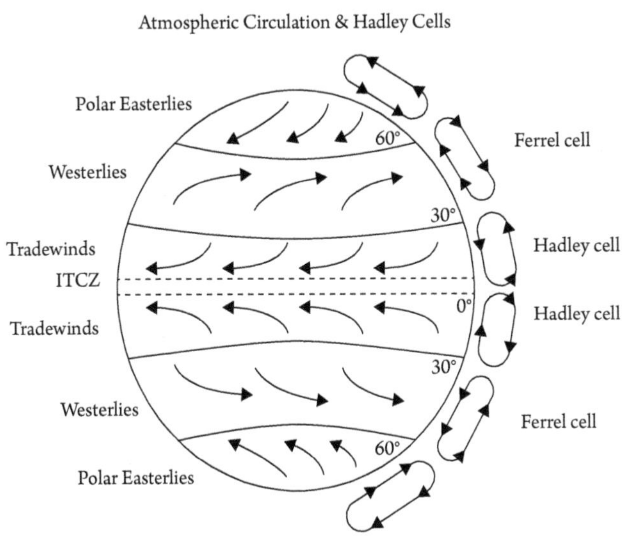

Figure 1.1 Schematic presentation of the cells in the atmospheric circulation. *Credit: Wikimedia Commons.*

There was widespread agreement that the simulation matched experience. As a result, the experiment was assumed to confirm that the system of differential equations, important parts of which had already been worked out by Bjerknes and Richardson, had been solved numerically. The quasi-empirical (because obtained in a simulation experiment) "proof" granted them the status that they continue to enjoy to this day of being *the* primitive equations of atmospheric dynamics. Hence, the simulation experiment raised them from the status of speculative modeling assumptions to the rank of an appropriate law-based description that now represented a core element of the theoretical modeling of atmospheric dynamics.

I want to stress this once again: none of these equations was inherently speculative; each one was securely founded in theoretical knowledge. The doubts to be found here centered on which compilation of equations would result in an adequate model. The formulation of a theoretical model was impeded or actually prevented by the circumstance that it was not possible to test the global interplay of what were locally plausible model assumptions.

Winsberg (2003) rightly considers this to indicate a general feature of simulations. They are indispensable in situations in which the global consequences of local assumptions are too complex to be determined through other (mathematical) means.[5] Numerical experiments changed the situation in meteorology exceptionally rapidly and led to the acceptance of the primitive equations proposed in the theoretical model. Above all, it became clear that model formulation and experimenting both work together in simulation. The leading theoretical meteorologist of the time, Edward Eady, recognized this immediately in a commentary on Phillips's experiment: "Numerical integrations of the kind Dr. Phillips has carried out give us a unique opportunity to study large-scale meteorology as an experimental science" (Eady, 1956, p. 356).

Hence, the experiment is used to derive statements from the model that can then be tested against reality. However, the aim is not just to confirm the theoretical model through the experiment. It is just as important

5. Winsberg provides telling case studies when he discusses artificial shock waves (2003) and artificial molecules—silogens (2006b).

to construct a discrete model and implement it as an executable program (Humphreys, 1991). Anybody who has ever done any programming knows all too well that a great deal of time and effort is required to move from the stage of a theoretical sketch to a compiled version of a program. And this does not include the time and effort spent detecting all the remaining "bugs." Therefore, the computer model cannot be described with pencil and paper. Everything that goes beyond the initial conceptual stages can be mastered only with an accompanying experimental program. Anyone wanting to track the program syntax as strictly and untiringly as the machine does would simply be unable to cope. I shall return to this aspect later when addressing epistemic opacity (chapter 4) and the limits of analysis (chapter 7).

EXPLORATIVE COOPERATION

Phillips's experiment involved two modeling steps that are conceptually distinct. The first was to formulate the theoretical model—in this case, a system of partial differential equations representing the laws of motion and fluid dynamics. The second step was to construct a discrete model to permit numerical processing with a computer. One can conceive the simulation model as a model of the theoretical model. Küppers and Lenhard (2005) therefore talk about simulation as "second-order modeling." Winsberg (1999) distinguishes an entire fine-scaled hierarchy of models extending across the space between the theoretical model and the executable simulation program. For my argument, however, it is the discreteness that is decisive. This second step is generally to be found in every simulation modeling, because each simulation model specifies a discrete generative mechanism, be this the technique of difference equations, cellular automata, artificial neural networks, or whatever. In the present case, a theoretical model served as the direct template, which is typical for difference equations because they refer back to differential equations. However, not every simulation model draws on a theoretical model as template. For some simulation approaches, an extremely weak theoretical basis will suffice.

Recent voices in both the philosophy of science and research are even using simulations to support the claim that we have come to the end of theory. Napoletani et al. (2011) have diagnosed that science will now proceed in an "agnostic" mode. Very similar theses can be found in various claims about big data and its ramifications. I find this precipitate. As I shall show, an intermediate course is appropriate: some, though not all, simulations are based on strong theoretically motivated assumptions, and it would be very one-sided to make any categorical judgment on this point. However, I also wish to stress that the example discussed here should not be viewed as the general case but, rather, the one producing the most powerful statements for my argument. This is because simulation-specific properties have to assert themselves clearly against the background of a strong theory. In that case, how does this approach differ from theoretically based modeling?

What characterizes simulation modeling is its close intermeshing with experimental practices. It is Phillips's experiment, and not the derivation from theory, that gave the equations their status as primitive. In other words, the observed agreement between measured and simulated flow patterns was the decisive factor, and the primitive equations did not form any independently validated element. Naturally, one could say that it would never have been possible to simulate the observed pattern if one had not started off with the correct theoretical model. This represents a common—and strong—argument regarding the validation of complex models: if you do not apply the correct theory, you will not get the pattern to fit.

This idea will be discussed in more detail in chapter 3, in which it will be seen that there is a broad range of situations and also cases in which a fit can be achieved via simulation even without referring to a theoretical approach.

Whatever the case, it is necessary to recognize that a major modeling effort has to be invested in the second stage of simulation. The selection of a space–time grid, the adjustment of the continuous dynamic, the implementation as an executable program—all these steps (or substeps) of modeling belong to a phase that often lasts for years. Throughout this phase, the behavior of the model has to be observed repeatedly, and there is no other way of doing this than experimentally. It is certainly true that recent error-analysis methods have made much progress (see, e.g., Fillion & Corless, 2014; Fillion, 2017) in estimating error bounds in systematic ways. Such methods, however, set mathematical conditions on how the simulation dynamics works (nonlinear systems of partial differential equations, for example, often do not behave very regularly). These conditions are not met in many practically relevant cases—general circulation models among them. The dynamic of the simulation models with their host of subprocesses fitted to the computer have to be tested through recourse to experimentation.[6]

This reveals a significant asymmetry: How can one determine whether the second modeling step was adequate? The only way to do this is to explore the overall behavior—that is, to judge the outcome of all the modeling steps. Nonetheless, such an approach is limited to those cases in which simulations reproduce certain phenomena with which one can compare them, as with atmospheric dynamics. Put briefly, if the global simulation proves to be satisfactory, the entire modeling would seem to be correct. However, if it fails—that is, if the simulated patterns diverge from the reality they are being compared with—it is not clear which steps in the modeling produced the error. If the simulation experiment has failed, if only instable patterns have emerged, or if the patterns have proved to be not very similar, then the theoretical model, the discrete model, and the implementation all have to be questioned. In general, local repair strategies have priority over a revision of the theoretical model—an approach that builds on the model's plasticity (see chapter 3).

Phillips had no alternative to the experiment when testing the adequacy of his modeling. Hence, exploring model behavior in a quasi-empirical way is a necessary condition that first makes modeling in any way possible—at least in complex cases such as these in which there is simply no other access to the model behavior. Then the theoretical model itself becomes the object of study, although it can be studied only in the context of additional assumptions such as grid size, parameter values,

6. To be clear, experimentation here refers to *simulation* experiments. The qualification of experiment is often dropped where the context clarifies whether ordinary or simulation experiment is meant (see also the introductory chapter).

and so forth. Simulation experiments serve to study the consequences of varying together both theoretical model and context.

The explorative and experimental aspect acquires its true significance only when simulations become not just numerical solutions of theoretical models. Naturally, this is already the case when there are no theoretical models of some reality. However, I wish to argue in favor of the stronger claim that even when theoretical models are available, the simulation model does not solve the theoretical equations in the way that solving equations is understood to function in (pure) mathematics. Solving equations numerically is— conceptually—very different from solving them analytically. Numerical solution thus is a concept that deserves philosophical attention. To support this claim, it will be helpful to look at the further development of the circulation models.

INSTABILITY AND INSTRUMENTAL COMPONENTS

Phillips's experiment is held to be the famous "first experiment" in meteorology. It produced the patterns of circulation, and was therefore a great success. However, it can simultaneously be seen as a failure because the dynamics failed to remain stable. After only a few simulated weeks, the internal energy ran out of control and the system "exploded"—the stable flow patterns that had been such a good fit dissolved into chaos. As Phillips himself reported laconically: "After 26 days, the field . . . became very irregular owing to large truncation errors, and is therefore not shown" (1956, p. 145). It was now precisely the abovementioned case that had occurred: The experiment failed to simultaneously confirm both the continuous and the discrete model. This failure led to a particular problem, because it was not clear what led to the observed malfunction and how one should carry out the necessary reconstruction. Phillips and his colleagues in the Charney group were fully aware that long-term stability is a decisive aspect. They soon decided that the cause must be a construction error in the discrete model. This, however, seemed to derive from the theoretical model whose equations seemed so very well confirmed. Hence, the group reasoned, the construction of the simulation model did not in fact follow from the theoretical model but, rather, contained unknown assumptions, and it was the failure to take these assumptions into account that nullified the approximation.

The first and obvious assumption was that the problem was to be found in the truncation procedure (as Phillips noted in the quotation given earlier). Such procedures are always present when mathematical representations that work with real numbers are turned into finite representations that allow only a certain number of digits. Computers are finite machines that work with stepwise algorithms. Hence, they require a transition from a continuous description to a discrete, step-by-step, and finite representation. When calculating with real numbers, errors are unavoidable. In principle, even a computationally more intensive refinement of the step length cannot overcome this problem, because the finer the space–time grid, the more arithmetic operations are needed, and this in turn results in more roundoff errors. Therefore, in practice, it is necessary to seek a pragmatic compromise between these two opposing sources of inaccuracy. Naturally, even a number rounded off to the eighth decimal place deviates from its real value, and even though this deviation is only minor, its effects can be major when many iterations are performed and the final state of the system is to be determined recursively. Quite correctly, the researchers recognized that finding an appropriate roundoff procedure is a major problem with far-reaching ramifications for simulation methods.[7]

Naturally, this was a significant concern in domains far beyond the bounds of meteorology, and it was basically not a problem exclusive to this field. It impacted on simulation in general. It now became clear that any way of replacing what mathematical tradition had felt to be a "natural" representation of the dynamic with the finite representation of a simulation model would raise a series of new problems.

7. Nicolas Fillion underlined this point in a personal communication (Feb 20, 2017): "When mesh size goes to zero, the floating point arithmetic becomes wrong because of overflow and underflow and other problems related to operating near the size of the machine epsilon. So, the real challenge is not finding a good roundoff (though it is one) procedure, but rather is that of finding a good adaptive mesh size that—at the same time—does not run into the over-/underflow problem, is accurate, and computationally efficient." Of course, Charney, Phillips, and their colleagues did not apply adaptive mesh sizes.

John von Neumann immediately recognized the significance of the stability problem for a broad field of applications for numerical methods. This vindicated his aforementioned appraisal that meteorology was a case with paradigmatic potential. In 1955, triggered by the stability problems in Phillips's experiment but before Phillips published his findings, von Neumann called a conference on the "Application of Numerical Integration Techniques to the Problem of the General Circulation" (Pfeffer, 1960). Its main topic was how to handle the unavoidable *truncation errors*—that is, to find a strategy that would make simulations both approximate *and* stable.

After several years of intensive research on this set of problems, it became apparent that it was not only the techniques of truncation but also the dynamic properties of the system itself that were decisive. Phillips (1959) discovered that a "nonlinear computational instability" was responsible for even very small errors building up into major deviations. The general thrust of efforts, including those of Phillips, was to develop roundoff procedures that would ensure that errors would average out before they could build up. The goal was to achieve the greatest possible perfection when calculating the theoretical model. These efforts were based on the view that simulation modeling was not an autonomous step but, rather, one serving to apply the calculating power of the computer in order to solve the continuous model. The research program aimed to ascertain the conditions under which the dynamic of a simulation model would deviate only slightly and in a controlled way from the theoretical model. However, this strategy miscarried—in my opinion—in a significant way: Computer modeling had got off on the wrong track, and only a conceptually new approach to the problem would lead to a breakthrough. This came with the insight that the task had to be defined on the basis of successfully imitating the phenomena and not on approximating the solution to the system of differential equations.

The breakthrough came from Akio Arakawa, a Japanese meteorologist and exceptional mathematician who developed a circulation model at the University of California, Los Angeles (UCLA). Right from the start, he was less worried about attaining a more exact solution through an improved code. He was far more interested in how to find a new kind of theoretical approach that would guarantee long-term stability without having to stick slavishly to the mathematical correctness of the approximation. Basically, Arakawa recognized that it would be necessary to dispense with the strict ideal of "solving" the primitive equations. If he could develop a simulation approach that adequately reproduces the temporal development of the atmospheric phenomena to be modeled and also does this in a stable way, then this simulation does not have to be a solution in the mathematical sense—not even in the limit. This was based on the theoretical insight that a solution in the limit may well be of little practical use. Even when a system of discrete differential equations converges with the continuous system in an increasingly fine space–time grid, it converges along a path on which the statistical properties of the discrete system differ greatly from those of the continuous system. This contains a general lesson: arguments that apply to the limiting case might be misleading if one does not actually work with the limiting case. Put briefly, Arakawa's standpoint maintained that imitating the phenomenon (under working conditions) has priority over solving the primitive equations (in the limit).

Naturally, this does not imply that any arbitrary mechanism can be applied to simulate a complex dynamic. Even Arakawa stuck closely to the primitive equations. Nonetheless, his approach granted simulation modeling a degree of autonomy from these equations. Arakawa did not start off by following the common practice of using the discrete Jacobi operator to describe the temporal development of the system; instead, he replaced this with his own special operator that then came to be known as the "Arakawa operator." Its derivation is mathematically very sophisticated, and the details do not need to be discussed here (see Arakawa, 1966, and the later reconstruction in Arakawa, 2000). What is decisive is that the Arakawa operator can be used to overcome the nonlinear instability so that the dynamic also attains long-term stability. He was even able to prove this property mathematically. This famous derivation gave him the reputation—which he still holds rightly to this day—of being the theoretically most adept mind in meteorology. This mathematical tour de force contained the real revolution, because it specified the stability as the most important criterion—and this was a property related to the *computer as instrument*.

To achieve stability, Arakawa introduced additional assumptions running partly counter to the theoretical laws on which the primitive equations were based. These included the assumption that kinetic energy is conserved in the atmosphere—which is certainly incorrect because part of it is transformed into heat through friction. For his colleagues, Arakawa's assumptions first seemed to be only mathematical gimmicks that were simply unacceptable if one wanted to engage in realistic modeling. As Arakawa (2000, p. 16) later recalled, the tone was "Why require conservation while nature does not conserve?" Most modelers were directing their efforts toward bringing the numerical algorithms ever closer to the primitive equations. From such a perspective, Arakawa's approach was incongruous, or at least dubious.

The history of physics is full of such examples in which contrafactual assumptions were introduced to make models manageable. The present case is also similar, only that Arakawa granted this freedom to the *simulation* model. Retrospectively, one could confirm that Arakawa was taking the path to give "physical" reasons for a simulation model by introducing instrumentally motivated artificial assumptions and wagering on the effects these assumptions would have in the experiment. The successful imitation of the phenomena in the simulation experiment was the benchmark that compensated for the lack of theoretical adequacy—conceived as the introduction of components that are artificial in relation to the theoretical model.

Consequentially, the theorist Arakawa had to fall back on the simulation experiment to underscore the relevance of his mathematical derivation. In this context, it is interesting to see that Arakawa had already developed and presented papers on his approach in 1962, but he only published it four years later after successfully implementing his operator in the model and carrying out simulations. At that time, almost everything had to be written in machine code, and he had to do most of this work himself. This was the only way to test whether the simulation was at all successful—above all, that it remained stable while simultaneously retaining the quality of the approximation without this being impaired by potentially false additional assumptions.

In my opinion, this was a decisive point: the discreteness of the model required artificial and also nonrepresentative elements in the simulation model whose dynamic effects could be determined only in a (computer) experiment. Simulation modeling is therefore dependent on experimenting. It was the success of the simulation experiment that finally led to the establishment of Arakawa's approach, with the initial controversy changing into broad support and even admiration. What started off as a questionable new approach was confirmed in the experiment and has now become so generally recognized as to be established as "Arakawa's computational trick" (see also the chapter with the same title in Weart, Jacob, & Cook, 2003).

Even though Arakawa's operator has been acknowledged since the 1960s as a great theoretical achievement—that is, as a mathematically motivated way to obtain an accurate *and* stable simulation—the majority of researchers were still concentrating on refining the truncation in order to produce the desired stable dynamic through an appropriate smoothing that would average the errors away. Once again, it was a simulation experiment that influenced the paths of modeling by delivering a decisive outcome. In 1978, Charney organized a large-scale benchmark simulation experiment in which three different models of the atmosphere were run parallel under the same starting conditions: those of Leith (Lawrence Livermore Laboratory), Smagorinsky (GFDL Princeton), and Arakawa (UCLA). Whereas the first two had implemented truncation methods, the UCLA model relied on Arakawa's operator. Phillips describes what happened:

> Three general circulation models . . . were used for parallel integrations of several weeks to determine the growth of small initial errors. Only Arakawa's model had the aperiodic behavior typical of the real atmosphere in extratropical latitudes, and his results were therefore used as a guide to predictability of the real atmosphere. This aperiodic behavior was possible because Arakawa's numerical system did not require the significant smoothing required by the other models, and it realistically represented the nonlinear transport of kinetic energy and vorticity in wave number space. (Phillips 2000, p. xxix)

Obtaining a numerical solution might require not focusing too closely on the "correct" analytical solution. This once again underlines the premise that the initial perspective viewing simulation as derived from a (mathematical or physical) theoretical model is a dead end. It then became necessary to apply truncation methods to achieve stability; this, in turn, impaired the long-term properties of the imitation. Proximity to the true solution does not ensure that the computed solution has the desired properties, such as stability and efficiency. This fact has become a key driver of modern numerical analysis since the 1980s (see Fillion & Corless, 2014). Arakawa's approach made it possible to produce a more "realistic" picture of atmospheric dynamics by using what were initially contraintuitive and artificial assumptions that were justified instrumentally rather than theoretically. Or to put it better, they were certainly motivated theoretically, but by the mathematically indicated numerical stability—in other words, an instrumental, computer- and simulation-related criterion that runs counter to theoretical adequacy in the sense of the physical primitive equations.

SIMULATION AND DISCRETE MODELS

The state of affairs can be abstracted from the case presented here and given in a more general form. Then one is dealing with a system C of partial differential equations and a system D of finite difference equations. Replacing C with D is justified essentially in that for refined nodes in the space–time grid represented by the parameter Δ, D transforms finally into C:

$$\lim D_\Delta = C \text{ for } \Delta \to 0 \qquad (*)$$

So far, so good. The relation (*) suggests that one can assume that when the grid is fine, the dynamic properties of D_Δ and C will also match in practical terms. Thereby, the simulation model D_Δ actually does deliver a numerical solution to C. However, on closer inspection, this conception does not hold, because in no way are all the dynamic properties already approximated on a fine grid. Put more precisely, there is no reliable estimate of what should be considered "fine enough." One could try to define "fine" by saying that a fine grid is one on which the properties are close together. Then determining the boundary at which a grid can be held to be fine is a question that can be clarified experimentally.

Viewing simulation as a numerical solution that is obtained from the theoretical model to be solved is misleading. The concept of numerical solution is richer. Hence, what are the benefits of such an enticing equation (*)? In practice, no grid is infinitely fine, and it is a hard, often too hard, task to determine something such as the distance between D_Δ and C for $\Delta > 0$. Even when numerical analysis can provide asymptotic bounds, the actual distance in practical (nonasymptotic) cases is a separate problem. In other words, the properties of model C, let us call them *Prop*(C), can, at times, differ decisively from $Prop(D_\Delta)$ for a given small $\Delta > 0$. Defining "small Δ" so that $Prop(D_\Delta)$ is, as desired, close to *Prop*(C), would beg the question.

This can be presented succinctly in formal notation. The vertical lines $|\cdot|$ should represent an appropriate measure. The choice of this measure will certainly depend on the specific circumstances: on the purposes and goals of the model such as the agreement between statistical variables and the patterns of global circulation. Although because of (*), the relation $|Prop(C), Prop(D_\Delta)| \to 0$ for $\Delta \to 0$, is to be expected, there is generally no estimation of $|Prop(C), Prop(D_\Delta)|$ for $\Delta > 0$. This statement has to be qualified. In not too irregular cases, asymptotic bounds can be obtained that tell how fast the distance becomes smaller as a function of Δ becoming smaller. But there is no guarantee that the properties will approach one another monotonously as the grid becomes finer. Hence, an asymptotic result need not provide the desired information about an actual run (of a climate model, say).[8] Hence, one is faced with a very open situation: neither is D determined by (*), nor is it determined unequivocally how Δ becomes (or should become) smaller, nor is $Prop(D_\Delta)$ known for any concrete D_Δ. Under some circumstances, a specific refining of the grid could even lead to a poorer approximation.

If one knows something about the dynamics of C and D, one can often derive mathematical statements on the approximation properties, such as an estimation of the speed at which the models approach each other (as the grid becomes finer). In complex cases, however, one generally has to manage without such statements. One important reason is the architecture of complex (plugged-together) simulation models that are a far cry

8. The growing field of backward error analysis (Fillion, 2017) tries to extract information about the path along which D_Δ comes closer to approximating C.

from the mathematical conditions numerical analysis calls for, especially when perturbation methods break down. (Chapter 7 provides a critical discussion of model architecture.) Viewing *D* as a simple numerical annex of *C* overlooks precisely the core area of simulation modeling.

If simulation modeling deals with finding a numerical solution, the notion of "solution" entails pragmatic and technical aspects (stability, efficiency, tractability) that distance it from the usual notion of a closed-form solution. This point is of major importance. In general, *D* provides a numerical solution for *C*, but not a conventional solution. The crucial property of *D* in order to numerically solve the equations of *C* is that *D* imitates a process. This is what happened in the example discussed here. The fact is that the properties of a model—the consequences that one has to accept during the course of its construction—can be observed in experiments and can be adjusted during the course of the cooperation between experimenting and modeling. This sounds as if simulation modeling is inevitably of an ad hoc nature—which it is not. While there are many ways in which simulation modelers can involve ad hoc procedures, there are also forms of rigorous testing and validation. The case discussed here gives an example. Chapter 7 will take a closer look at the problems with validation.

One crucial point in the methods of simulation modeling is that they transform the comparison of the phenomena and the data with the model into a part of the model-construction process. This can be understood by looking at the adjustment of parameters. Practically, almost every model contains a number of parameters. The case of simulation is particularly unique, because the specification of the parameters often does not belong to the initial or boundary conditions. The parameters applied typically in simulations are not determined "from outside" but, rather, "from inside" through comparisons with the results of numerical experiments—that is, through a kind of reverse logic. An appropriate parameter value is then the one that leads to useful results within the framework of the simulation dynamic being tested. As a result, even black-box parameters permitting no clear interpretation can be applied successfully.[9]

9. Explorative cooperation of exactly this kind is also essential in thermodynamic engineering as Hasse and Lenhard (2017) point out when discussing "the role of adjustable parameters in simulation modeling."

Let us look at this process in a strongly idealized form. Take a complete host of parameters $p := \{p_1, p_2, \ldots\}$, and let us call them adjustors for the discrete model D: $D = D(p)$. $Prop(D(p))$ is conceived as a function of $p = \{p_1, p_2, \ldots\}$. The goal is to vary p in such a way that the observed behavior (in the numerical experiment) approaches the phenomena being modeled (Ph). The task is:

Find p, so that $|Prop(D(p)), Prop(Ph)|$ is small.

The formal notation of the goals suggests more of a quantitative solution than is generally possible. But even a more qualitative judgment by experts does not demolish the argument. This is because the appropriate model specification, that is, the concrete parameter values p, is found within a process of explorative cooperation between modeling and experimenting. That which is achieved traditionally through approximation to the theoretically valid conditions can now be achieved by comparing the properties of the model with the phenomena.

AUTONOMY AND ARTIFICIALITY

I shall briefly summarize the content of this chapter: Simulation modeling can be conceived as an interplay between theoretical and instrumental aspects. This tends to remain concealed by the presentation of the discrete model D as the solution. This suggests a sovereignty of definition on the side of the (traditional) theoretical model C, although this actually exists in only a weak sense. "Instrumental" here is an interesting notion that does not match the established notion of instrumentalism and its (in)famous opposition to realism. This controversy has lost nearly all its productive power (if it ever had any). An updated form of instrumentalism that is related to simulation will be discussed in the closing chapter 8.

In some ways, the method of finite differences represents a theory-intensive case, insofar as a strong theoretical model is present that may well be lacking in many other simulation approaches. Other cases of a similar kind are molecular modeling or thermodynamics engineering. It is only in the extreme case that talking about the solution in any way develops its force of attraction. Now it is precisely this case that has received much attention in the philosophy of science, in which it is viewed partly as an autonomous category and partly as a generic case. However, I consider this to be inappropriate because it loses sight of the common properties of simulation modeling. Consequently, my main motivation for addressing such a case in detail here is the fact that it is located at the theoretical end of the spectrum of simulation models. The results obtained here on the application of experiments and artificial, instrumental elements apply a fortiori for other modeling techniques.

Put briefly, there is a strong need to modify the concept of simulation as a numerical solution of a theoretical model. Simulation models, as well as numerical solutions, depend on more factors. It is generally not the case that a previously formulated exact theoretical model needs to be protected from the harmful influences of heuristic and explorative elements, as Stephan Hartmann (1996, p. 87) points out, in order to counter the tendency to devalue the influence of simulations. In fact, simulation modeling faces a task in which success depends essentially on the interplay of simulation experimenting and modeling—that which I have called *explorative cooperation.*

This closely intermeshed connection confounds the differentiation between a *context of discovery* and a *context of justification,* because comparisons are performed continuously during the model-formulation phase. Here, I agree with Friedrich Steinle (2005), who has worked out convincingly that the development of the modern sciences has always been accompanied by a major share of (ordinary) explorative experiments. And Steinle (2002) comes to the conclusion that this questions the differentiation of contexts. This explorative share becomes, as I have shown, a central element in the methodology of simulation.

When following an explorative approach, performance is the crux of the matter. In order to optimize performance, modeling decisively includes not only theoretical ideas but also the phenomena and the data. This does not just apply to the finite differences but also refers to generative mechanisms completely in general. Where such mechanisms are of an artificial nature and do not represent the phenomena to be modeled, it is performance that remains the criterion for successful modeling. However, performance can be determined only in the experiment.

For theoretical mathematics, the study of discrete generative mechanisms is a relatively new topic. Numerical analysis makes progress in covering classes of mechanisms by taking a predictive approach. When these mechanisms are not defined mathematically but, rather, result from piecewise and diverse modeling processes, predictions become problematic. Then, no powerfully predictive theory is available, and the model dynamic tested experimentally on the concrete object has to do the main work.

For a long time, conceiving the performance of complex models theoretically was a major goal in theoretical informatics. Mathematization seemed to be the royal path to establish informatics as a theoretical science. In a series of studies, Michael Mahoney (e.g., 1992) has shown how unconvincing the results of these efforts have been. One can view this as confirming the essential role of explorative cooperation. Hence, essential characteristics of simulation modeling are that experimenting and modeling are closely interlinked and thereby both also necessitate and permit the application of artificial or instrumental components in the service of performance.[10]

Hence, simulation models open up a specific perspective within the current debate on the role of models in the philosophy of science. The type of theoretical model labeled C in the discussion earlier has long dominated model formulation in mathematics and the natural sciences,

10. Indeed, it is easy to add further examples to the case of meteorology I have chosen here. Completely in this sense, Winsberg (2003) introduced the example of "shock waves" in which artificial wave fronts are added to compensate for discretization effects). This category also includes the parameterizations of the climate model that present processes at a grid node that actually occur in the area between the grid nodes—although this is not included in the simulation model. A series of philosophical works are available on this topic that focus particularly on the context of climatology, such as Gramelsberger and Feichter (2011) or Parker (2006). Ulrich Krohs (2008) describes an example from biology in which the simulation modeling proceeds from a theoretical model but then resorts to measures that are labeled instrumental here.

and has also been at the center of philosophical attention since Hertz and Maxwell— who had important forerunners such as, among others, the Bernoullis and Newton. The contemporary debate is based initially on the insight that one cannot proceed from a theory to applications without models. It then goes on to acquire a controversial tone through the theses of Nancy Cartwright and others that theories, or, to put it better, theoretical laws, are of little help when it comes to application issues.[11] Therefore, according to Cartwright, first, theoretical laws are as much as useless in any relatively complex application context; and second, models are guided far more by the phenomena and the data (termed *phenomenological* models). Margaret Morrison proposes a kind of midpoint in "models as autonomous mediators" (Morrison, 1999; see also further chapters in Morgan & Morrison (1999).[12] Accordingly, models are determined neither by the theory nor by the data. It is far more the case that they adopt their own autonomous role that mediates between the two.

The findings in the case of simulation models confirm the line taken by Morgan and Morrison: the formulation of a (discrete) simulation model is not just a mere derivation from a theoretical blueprint. This raises the autonomy aspect again—in this case, between the theoretical model and the simulation model. Arakawa's work serves as an example of how success can depend on conceiving simulation modeling as an autonomous task. At the same time, however, it has to be remembered that the theory—the theoretical model of the atmospheric primitive equations—nonetheless took an extremely important position in the simulation modeling. And finally, it becomes clear how the phenomena and data exert a direct influence on the simulation model in the interplay between experimenting and modeling. This makes it completely

11. Such a standpoint, by the way, has frequently been defended in historical studies of science and technology.

12. The philosophical discussion on models is now so widespread that I may forgo a detailed presentation. Suppes (1961) can be seen as a pioneer here; Cartwright (1983), Giere (1988), Hacking (1983), and Morgan and Morrison (1999) are standard references for the recent debate.

acceptable to view simulation models as autonomous mediators. Beyond confirming this general thesis, simulation modeling adds a new aspect to the model debate, because the mediation performance depends on the particular properties of the discrete models—and the instrumental ways in which they are handled.

2

Visualization and Interaction

Mathematical models can be presented in quite different ways. Even at first glance, simulations stand out because they often take a visualized form. Without doubt, this visualized form of presentation is one of their major characteristics. However, the point about computer experiments addressed in the previous chapter also applies to visualizations: neither should be taken as the uniquely determining feature. Nonetheless, one aspect of visualization makes the greatest contribution to the characterization of simulation modeling—namely, the way in which visualizations are embedded in the modeling process.

Just as numerical experiments can draw on a long tradition of empirical methods in the sciences, visualizations connect to a rich field of pictorial (re)presentations. There is no inherent logic of development toward computer visualization, although a couple of features make visualizations stand out as a particular kind of representation; one example is the amount of control that is feasible in the production of simulation images, including the variation of colors, perspectives, textures, and so forth.

In visual studies, simulation is found mainly in the context of *artificial* representation.[1] In the philosophy of science, one can find a closely related debate over the ways in which scientific models represent.[2] Hence, this chapter will concentrate on simulation models. It will consider neither

1. Reifenrath (1999) offers a typical instance of the position that prevails in the history of art and in visual studies. He treats artificiality as the main feature of simulations.
2. Here, I would like to refer to the contributions of Bailer-Jones (2003) and Suarez (2004). The former discerns a series of ways in which models represent, whereas the latter denies that models represent at all (in a certain sense, of course).

pictorial representation in general nor a philosophically adequate concept of representation in particular. Instead, the argument will be much more specific and address the role of visualization in the process of simulation modeling. In short, my claim is as follows: from a methodological perspective, visualization supports the exploratory and iterative mode of modeling; from an epistemic perspective, it assigns a special role to judgment.

Of central importance for both perspectives is that visualizations offer opportunities to interact with models. Researchers may vary representations on screen or highlight particular aspects of processes. Such interactions combine the impressive human powers of comprehension in the visual dimension with the computational capacities of digital machines. A visual presentation fosters the interplay between experimental approach and instrumentalist assumptions discussed in chapter 1. Visualization can be decisive when adapting such assumptions in line with the performance of the simulated system. This form of representation makes it possible to compare patterns; and this comparison can then serve as a criterion for the adequacy of models even when the adequacy cannot be derived (theoretically) from the assumptions that have been made. In this case, visualization can play an essential role in the process of model building.

Prima facie, the wish to include experimenting and exploring in the process of modeling and to do this in an effective way presuppose high standards of instrumental quality. It may be necessary to vary model assumptions and to observe and control their impact on model behavior even when there is no analytical (theoretical) grasp of these effects.

FUNCTIONS OF VISUALIZATION

Two features particularly motivate the focus on visualization—namely, the *intuitive accessibility* (*Anschaulichkeit*) of the presentation and its *process character*. Both are closely interrelated.[3] When a presentation

3. Fritz Rohrlich (1991) was referring to both at once when he underlined that simulations are "dynamically *anschaulich*."

is intuitively accessible, researchers have the opportunity to grasp important aspects of the model dynamics and to refer to them directly on screen—at least in the ideal situation. Such access can provide important hints when having to judge the adequacy of a simulation. In Phillips's numerical experiment discussed in chapter 1, the simulated dynamics of the atmosphere appeared to be so good, because the simulated patterns were such a good match with the observed ones. Problems of that kind, in which patterns are compared with each other, occur frequently in the context of simulations, and they can be tackled especially well on a visually intuitive basis.

The reasoning about the qualities of visual representations does not invoke the computer in any particular way. This machine, however, does enlarge the scope of applications considerably. Two complementary aspects interlock: On the one hand, enormous amounts of data become available that are generated by automatized—and this means mostly computerized—procedures.[4] On the other hand, even vast amounts of data can (in favorable cases) be condensed into telling visualizations. In many cases, there is no alternative representation at hand to capture the data.[5] Generally, only machines can provide the necessary computational power to produce such representations. To determine how visualization functions as a part of simulation modeling, it is helpful to return to the process character just mentioned.

If the pictorial presentation takes a dynamic form, one speaks of a process-related presentation or animation. The process character has a special relationship to simulations. Stephan Hartmann, for instance, perceives this as the defining property of the latter (Hartmann, 1996, cited in Humphreys, 2004). The process character of visualization then corresponds to the process character of simulation. There can be no doubt that this marks an important aspect—although, in my opinion, it would be excessive to demand that a simulation should always imitate a process. A Monte Carlo simulation, for instance, can simulate an integral that itself can only with some effort be described as a process. I would like to

4. Although the phenomenon is widely acknowledged, there is still no commonly established name for this. *Big data* or *data-driven research* are prominent labels.

5. A new science or a scientific art of visualization is starting to form (see Chen & Floridi, 2013).

take up Hartmann's and Humphreys's point, but instead of referring to the target that becomes simulated, I would like to refer to the process of model building. There I shall show how controlled variation of visualizations provides an essential means for constructing and adapting simulation models.

The particularly intuitive quality of dynamic visualizations is a double-edged sword. Along the one edge, such visualizations can lead to success when complicated relations are presented in a way that makes them graspable. This may be the case even when the presentation greatly reduces the amount of information in the data. Especially computerized procedures generate masses of data that are hard to manage. The Large Hadron Collider (LHC) at CERN (European Organization for Nuclear Research) spews out such a vast amount of data that only a fraction of it can be stored (see Merz, 2006, for a study on the roles of simulation in the planning of the LHC). More examples will follow further in this chapter. Whatever the case, it is obvious that complex dynamics and vast amounts of data are hard to comprehend other than via computer visualizations. They work as a technology-based antidote to the technology-based accumulation of complexity and quantity.

Along the other edge of the sword, both the dynamic nature and the richness of detail that visualizations often portray can easily lead to a fallacy—one might say to the aesthetic fallacy—that one is observing presentations of empirically grounded processes. Hence, this other edge of the sword is that simulated systems have an appeal that makes them look like real systems. Hartmann (1996, p. 86), for instance, identifies this as one of the great dangers of simulation. The artificially realistic appearance on the screen can be (mis)taken for a strong indicator that the simulation represents real events. A contributing factor is the commercial nature of today's software packages, because visualization programs and graphical tools are developed and distributed as commercial products, and researchers implement them as ready-made modules, or black boxes. Software packages are distributed broadly over a wide range of scientific and technical fields, and more and more research institutions have their own professional visualization unit. Indeed, the aesthetic qualities of simulations in technical sciences are often very similar to those of computer games.

During the modeling process, it is tempting to include more and more detail, because this creates a more and more realistic visual appearance. However, such a course of action will not produce a true representation; rather, it offers an illusory one. Nevertheless, this is an often-noted, though equally often criticized, tendency—a fact that indicates the attractive force of powerful visualizations.[6] The resulting problem combines facets from the philosophy of science, epistemology, and aesthetics and has led to the development of an interdisciplinary field encompassing science studies, philosophy, and the history of art that studies the form of representation in simulations—that is, the pictorial quality of computer visualizations.[7]

Although the perfection of visualizations suggests direct visibility, they are constructed to a high degree and require a host of operations, decisions, and modeling steps to appear in the way they do. In fact, the dynamic and intuitive appearance is far from direct—it is highly mediated, and this is (in theory) a commonly accepted fact. The apparent palpability is especially remarkable in those instances in which there is no object that is being represented. Computer images often lack an "original" counterpart. Hans-Jörg Rheinberger (2001) treats visualizations as cases in which the object and the medium of presentation are in part identical. Although this is correct, little attention is paid to the function of visualizations in the process of modeling.[8] This function is the concern of the present chapter.

UNDERSTANDING COMPLEX DYNAMICS

A famous and early example of the investigation of complex systems is the work of the meteorologist and physicist Edward Lorenz. He studied a relatively simple dynamic system—again, a system of differential equations—and found that this system exhibits remarkable and unexpected behavior.

6. One example is the proximity between visualization and public relations and advertising in nanoscience; see the contributions in Baird, Nordmann, and Schummer (2004).
7. See Bredekamp (2003), Heintz and Huber (2001), Ihde (2006), Johnson (2006a), Jones and Galison (1998), and Lynch and Woolgar (1988).
8. Exceptions can be found in the work of Winsberg (1999), Martz and Francoeur (1997), Ramsey (2007), and Gramelsberger (2010).

A merely minor change in the initial conditions could lead, after passing through a completely deterministic dynamic, to a very big change in the overall long-term behavior. This is a property of the system that has aptly been called "deterministic chaos" (see Lorenz, 1967, 1993). It is a deterministic system; that is, as soon as the initial conditions are fixed, the whole trajectory is determined. At the same time, it is unpredictable in the sense that even the slightest variation in the initial conditions—in practice, one can never determine conditions without some variation—opens up the complete range of potential long-term behavior.

Lorenz's example is famous and has become a paradigm for the later development of theories of chaos and of complex dynamic systems. His original system, however, was not designed intentionally to show these somewhat exotic mathematical properties. Instead, it sprang from Lorenz's work in meteorological modeling. Therefore, the results were of direct relevance for meteorology.[9] The "butterfly effect" illustrates nonlinearity well: even minor events such as the flap of a butterfly's wing can have major effects in nonlinear systems such as the long-term weather—at least in principle. All this affects dynamic systems theory, a discipline that is formulated in the mathematical language of differential equations. The point is that one can detect the erratic behavior with the help of analytical methods, but it is not possible to understand what actually happens in such complex systems. How should one imagine a system that would react to the smallest of jiggles and be attracted to very different states? This is the point at which computer simulations and, in particular, visualizations come onto the stage.

The puzzling character disappeared only when computer-generated images, such as the one displayed in figure 2.1, made the Lorenz attractor (in phase space) intuitively accessible. It was the visual presentation that led to the development of a theory of complex systems. One needed something that would convince researchers that there were stable objects out there that would be accessible to systematic investigation. The image

9. After initial hesitation, Lorenz's considerations have been commonly accepted as a major breakthrough (cf. the historical outline in Weart, Jacob, & Cook [2003], and the previously mentioned works by Lorenz himself).

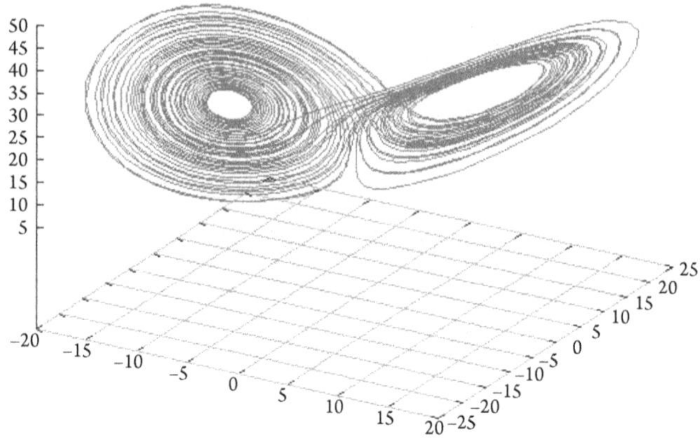

Figure 2.1 3D plot of the Lorenz attractor; 2900 points have been computed with the Runge–Kutta method and fixed steps. *Credit: Wikimedia Commons.*

shows how an initially close neighborhood of trajectories can lead to different states.

Figure 2.1 is a computer visualization in an essential sense, as will become clear from looking at the way it is constructed. To create images such as this one, for every point (grid or pixel), the behavior of the system is computed under the conditions (parameter values) particular to that point. In this way, a system is sounded out; the isolated results for very many points are gathered and then assembled into a single image. Only then does an image of the overall dynamics emerge. It looks like a continuous drawing, but it is constructed by assembling very many single results. Such a procedure depends on the computational power of digital machines. Why would one trust such visualization? Even if each single point in phase space is not positioned where it should be, and even if the solution is exponentially sensitive to perturbation of the initial conditions, the shape and dimension of the attractor is very robust under the same perturbation.[10] Intuitively accessible visualizations can make it possible to gain a view of complex dynamics even when the mutual interplay of the assumptions that went into the models is not well understood. The

10. I owe this point to Nicolas Fillion.

enormous computational power allows an exploratory mode to work with visualizations in the example given: exploration of phase space. And visualizations, in turn, open up intuitive accesses to complex system behavior.

This consideration elucidates the heading of this section, "Understanding Complex Dynamics." At first glance, complexity seems to rule out understanding of the dynamics almost *by definition*. I would like to maintain that simulation modeling, and visualization in particular, can provide an adequate means to investigate complex dynamics. They are no panacea, however; they are themselves limited by complexity. This limitation will be discussed in chapter 7.

In short, the visual presentation of strange attractors and fractal sets was a necessary element in order to transform the field of complex systems from a collection of curious phenomena into a research program. This program is full of visualizations: the volumes of Peitgen and colleagues (Peitgen & Richter, 1986, 1988) are mathematical monographs with the qualities of illustrated books. These so seductive presentations harbor a problem that is typical for simulations and has been discussed already in chapter 1: If it is assumed that the computer-generated image gives a correct impression of the behavior of the dynamic system, then even the slightest changes will matter. Consequently, the whole computer visualization will become questionable, because it is based unavoidably on discretization; that is, it changes the system at least a bit. This creates a paradoxical situation: How can a grid-based image display a pattern that makes it plausible for the exact particularities of the grid to matter?

In fact, there is reason for skepticism. The appearance of visualized images can be manipulated very flexibly to adapt color, to smooth textures, and so forth. However, it is exactly the manipulability that raises the question whether what are observed in the visualization are, in fact, instrumental artifacts. "Artifacts" here mean features that do not emerge from the model dynamics. The dynamics of the phenomena to be modeled are a different matter altogether. In the example of chaos theory, the visualized fractals and strange attractors had been met with skepticism until it gradually became clear that their essential properties have a robust nature.

Again, this ends up with the double-edged nature of visualization. It is a philosophical problem that emerges regularly when new instruments start to come into use. Then, enlarged instrumental capacities and opportunities are accompanied by a healthy skepticism regarding whether one is dealing with real phenomena or with artifacts created by the instrument. The philosopher Hans Blumenberg has formulated this problem as an antinomy:

> Galilei's reaching for the telescope includes an antinomy. He renders the invisible visible, and thereby intends to lend evidence to Copernicus' conviction; but by doing so, he commits himself to the risk of visibility as the final instance of truth; by employing the telescope, however, to establish visibility, he simultaneously breaks with the postulate of visibility held by traditional astronomy. Consequently, he cannot escape the suspicion that the technically mediated visibility, as advanced as it may be, is merely a contingent fact bound to conditions extrinsic to the object under study. (Blumenberg, 1965, p. 21; my translation)

This problem occurs even more acutely in the context of simulation and its uses of visualization, because the technological mediation happens in such adaptable and variable ways.

POWER AND SPEED

The pervasiveness of visualization is a relatively recent phenomenon. Whereas computer simulation originated more or less simultaneously with the electronic digital computer, visualization needed a high level of computing power and graphics technology to attain intuitive and dynamic qualities. The theory of complex dynamic systems, for instance, gained momentum only in the 1980s—Lorenz's much older results from the 1960s were, so to speak, too early for the state of the technology. Another instance is the distribution of computer-aided design that started around the same time.[11] An important motor of technological development definitely was and still is the game industry, with its strong demand for realistic appearances.

11. Johnson (2004) analyzes the pioneering phase by discussing a case from aircraft design.

In the context of simulation modeling, the speed of visualization acquires a particular role. The visual depiction of model results has a long tradition; diagrams belong to the inventory of every mathematically inspired science. But it is only with sufficient computational power that one can attain visualizations with a dynamic character. First and foremost, this means modifications of the model have to be transformed quickly into modifications of the visual presentation—ideally in real time. It is only with this condition in place that visualization can play its part in exploratory modeling. Trial and error, in short, is no option when the answer to the model demands a protracted calculation.

When the instruments offer more possibilities, visualization-based interaction with the model can play an important role: manipulation and assessment of the model happen in rapid succession. To speak of "playing with models" (Dowling 1999) not only involves an experimental aspect but also requires instrumental possibilities that allow such "play." Humphreys (2004) has pointed out that what matters is realizability in practice rather than "in principle." Simulation is an instrument that has to work in practical contexts. Hence, algorithms that yield an adequate result in principle are relevant only if they do so in an acceptable time. Humphreys has condensed this consideration into the slogan "speed matters." I agree and would like to add that this slogan is also relevant in the present context of interactive visualization. The latter realizes a feedback loop between behavior of the model and modification by the modeler, and this loop has to attain a certain speed to become practicable from the perspective of the modeler. In this sense, simulation modeling as a scientific procedure is dependent on technology.

Here's a historical remark in this context: simulation modeling appears to be similar to the program of cybernetics (Ashby, 1957; Wiener, 1948). Cybernetics had claimed to aim at effective manipulation via ongoing observation and control—without relying on an analytical understanding of the dynamics. This goal had elevated the feedback between humans and machines to a universal principle. This program, however, was tied to analogue machines such as Vannevar Bush's Differential Analyzer (cf. David Mindell, 2002, for an excellent historical study). Thus, somewhat ironically, the idea of feedback mechanisms has acquired significance by making a detour—namely, via the employment of visualizations that only appear to be analogue, whereas their essential adaptability is based on digital technology.

FORMATION OF GALAXIES

Visualizations can be essential supplements to models even when these models rest on a profoundly theoretical nature. The following example starts with an elaborated (though not validated) theory of how spiral galaxies form. A model of cellular automata was developed on the basis of this theory, and this model attracted attention in the early 1990s because it brought the simulation-specific approaches of cellular automata (CA) to the long-established and theoretically grounded field of astronomy—although CA had the dubious reputation of running counter to traditional theoretical accounts. I shall show how simulation experiments, visualizations, and theoretical modeling are interrelated.

The CA model serving as my example stems from Seiden and Schulman (1990), who investigated the development of spiral galaxies. They were able to show in simulations that a percolation model of galaxy structure based on the theory of "stochastic self-propagating star formation" (SSPSF) can explain the characteristic spiral patterns. The SSPSF theory was formulated analytically but could be assessed only within narrow limits. Their strategy was, first, to translate the theory into a simulation model. This step is similar to the discretization in the case of atmospheric modeling. Second, they wanted to use the results of the simulation to test and validate SSPSF against known phenomena of galaxy formation.

Seiden and Schulman constructed a CA model that started with a cell structure similar to a dartboard (see figure 2.2). Each of these cells is either occupied (black) or not (white). After each time step, the state of each cell is recalculated from the states of the neighboring cells. The initial state of this automaton seems to lack any resemblance to a galaxy. Hence, it is important to adjust the interactions with neighbors so that the final state will acquire a strong resemblance to a galaxy—which, stated in crude terms, consists of interacting clumps of stars. This can succeed only when the interactions between neighbors can be steered into quite

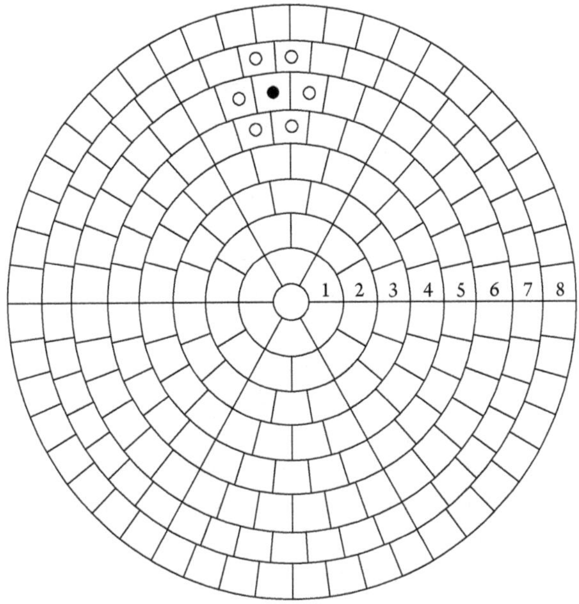

Figure 2.2 Schema of cells on concentric rings. *Credit: From Seiden and Schulman (1990). With kind permission of Taylor and Francis.*

different dynamics by adjusting parameters. The plasticity of simulation models will be discussed separately in chapter 3.

The specification of how neighbors interact is motivated but not determined theoretically. Therefore the visualized results of simulations build a necessary intermediate step in the test of whether the theoretically motivated assumptions in their dynamic interplay indeed lead to plausible results. The *primitive equations* of circulation models had a comparable status: each equation was validated theoretically. But then, the question was: Do the equations, as a dynamic system, generate the interesting phenomena? Only Phillips's experiment showed that the model was promising, and similarly, only the simulation documented by Seiden and Schulman showed that the theoretical model was adequate:

> Without the dramatic images produced by computer simulations of the galaxy model, the pencil-and-paper theorist would have had

> a hard time establishing that the model produces spiral-arm morphology or anticipating other astrophysical points of agreement. However, once these features are established, their underlying rationale can be appreciated with the aid of analytic techniques. (Seiden & Schulman, 1990, p. 31)

The Lorenz case also fits in with this. Without visualizations, strange attractors were mainly strange; whereas with visualizations, the theory of complex dynamic systems, gained leverage.

Of course, everything depends on the exact specification of the dynamics. Translating various physical processes into the cellular spiral, controlling the rotation of the disc, refining the cellular grid during the simulation, and so forth are all subtle issues. The architecture and schema of parameters often follow theoretical considerations, whereas the assignment of values takes place in interaction with simulation experiments. These parameters and their values influence the result decisively because they codetermine the form of the (simulated) galaxy. Which parameter values are adequate, however, can be learned only in a "backward" logic from the resulting simulation image. I have already presented this logic in the atmosphere example. In the present case, however, the backward logic is not a makeshift. Instead, it is employed systematically right from the start.

At decisive points in their model, Seiden and Schulman (1990) provide room for further adjustments. They do not have any other option, because one cannot determine the parameter values of CA by theoretical considerations alone. The parameters are defined in the realm of a discrete computer model, not in the continuous world of stochastic theory. Hence, they are assigned during the course of simulation modeling—namely, by comparisons with simulation experiments and their visual results. For instance, how big the cells of the disc should be determined quasi-empirically: "It will turn out in our simulations that their linear dimension is about 200 *pc*" (p. 20). The time step of 10^7 years per (CA) step is legitimized in the same way: "For this assumption, just as for the subsuming of many physical properties into a single parameter p, the test will be compatibility with observation" (p. 20).

This creates the following situation: starting from the not yet confirmed theory (SSPSF), a discrete version of it is constructed and implemented as a CA model with a couple of adaptable control parameters. It would not be possible to derive adequate values for these parameter values from the theory, because they do not appear in it. But how can one in any way find parameter values that make the dynamics of the CA model sufficiently close to the empirical phenomena? The key here is visualization.

First, adequate patterns have to be generated over the course of iterated experiments with varying parameter values; only then will the theoretical model be accepted. The *adaptation* of the simulation model to the phenomena and the theoretical *foundation* of this model are not mutually exclusive. On the contrary, simulation modeling can succeed only through their cooperation. Arguably, Seiden and Schulman would not have been able to build a CA model that would generate adequate patterns of galaxies without the theoretical basis.

The space of possibilities is far too large for unguided fumbling. Rather, the similarity of the generated patterns requires plasticity of the model and visual comparison; but it also indicates the adequacy of the theoretical assumptions, because inadequate ones would hardly lead to a successful adaptation.[12] It is also true, however, that without the backward-logical employment of visualizations, the model could not have been specified. On the level of neighboring cells—that is, in the discrete model— there is no argument in favor of particular parameter values. Any such argument would have to be of a global nature, but the global dynamic is accessible only from the visual results of the simulation.

In the case of the galaxies, the goal was for the simulation model to reproduce previously observed forms of galaxies. The criterion for success was the visualization of the CA model after many iterations. It gives a fine-grained image in which occupied cells have finally shrunk to black pixels. Depending on the parameter assignments, these images

12. Such considerations rest, however, on the assumption that practical success in such complicated matters relies on adequate theoretical guidance.

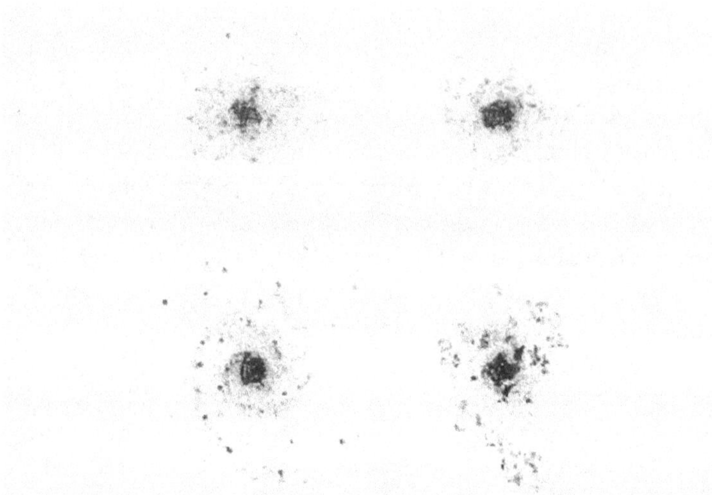

Figure 2.3 Two spiral galaxies, NGC 7793 (above) and NGC 628 (below); photographs on the left and simulations on the right. I have reversed black and white to make the outcomes more visible. *Credit: From Seiden and Schulman (1990, p. 52). With kind permission of Taylor and Francis.*

resemble photographic images made with the help of telescopes (figure 2.3).

This resemblance is the decisive criterion for Seiden and Schulman, because it shows the similarity between their CA model and stars. There is no theoretical measure for judging similarity—what counts are the images. The visual agreement documents that the structure of the CA model together with parameter adaptations can generate the forms of spiral galaxies.

SOME CONCLUSIONS

The method of CA has been chosen deliberately. Visualizations occur in one way or another in virtually all simulation approaches. In CA, they are connected particularly to the method, because CA models the dynamics by local interactions between discrete cells.[13] This leads to a new type of mathematical syntax that calls for a new type of mathematical modeling. On the one hand, this syntax is ideally suited to the capacities of the computer—namely, to calculate a great number of simple interactions iteratively, cell by cell. On the other hand, what range of objects and dynamics it makes sense to model is far from obvious. It turned out that simple neighbor interactions, using adequate scaling and adjustments, are extremely adaptable.[14] I shall discuss the plasticity of CA and other simulation approaches in the next chapter.

However, two points have to be admitted: First, CA models do not require the computer *in principle*. Like everything performed by a computer, a CA dynamic could be worked out step by step with pencil and paper. The simplicity was even the initial motive for introducing CA. Knobloch (2009) describes aptly how the economist and Nobel Prize winner Thomas Schelling switched from calculations by hand to a computer model while thinking about his famous nearest neighbor models for social segregation. What may be right *in principle* cannot necessarily be transferred to more extensive cases that are, *in practice,* not tractable without the computer. The latter sort of case is the more interesting one in many applied contexts. Second, not all cases of simulation involve feedback between a visual result and modeling assumptions or parameter assignments. Wolfram Media's commercially highly successful software package "Mathematica" uses CA presentations to process mathematical equations efficiently that have been typed in symbolically.

A number of philosophers and scientists have recognized cellular automata as the core of what is novel in simulations.[15] This is where a new mathematical syntax that abandons differential equations and global viewpoints meets visualization as an apparently direct mode of assessment

13. The approach goes back to the proposal that S. Ulam made to his colleague J. von Neumann, in which he suggested starting the theory of self-replicating automata with oversimplified architecture and syntax (see Aspray, 1990).

14. For instance, agent-based simulations are a member of the CA family and have led to an autonomous discipline in sociology (see Ahrweiler & Gilbert, 1998; Conte, Hegselmann, & Pietro, 1997; Epstein & Axtell, 1996; Hegselmann, 1996).

15. This group includes Keller (2003), Rohrlich (1991), and also Wolfram (2002) with his ambitious claim that CA provide the basis for a "new kind of science."

that can do without such complications as an analytical description of equilibrium states. Admittedly, these points represent palpable differences compared to other simulation methods. In particular, the finite difference methods discussed in chapter 1 build on traditional mathematical modeling, whereas CA follows a pathway that is oriented toward the computer right from the start. Nevertheless, I would like to relativize the claim for radical dissimilarity. It is more the case that CA simulations fit into the conception of a type of mathematical modeling that has been introduced in chapter 1.

The great significance of visualization should not lead one to neglect the fact that the procedure shows structural similarities to finite difference simulations. The quasi- experimental approach of exploratory cooperation is at work in both cases. Model parameters p are calibrated by observations in theoretical model experiments so that $Prop(D[p])$—that is, the dynamic properties of the discrete model—fit the phenomena. In short, visualization fosters the exploratory mode. Of course, different cases display different gradations regarding how influential the theoretical basis is and how extensively the backward logic (calibration) is employed. I would like to maintain that both aspects are not mutually exclusive. The galaxy example has shown that. Instrumental, exploratory, and theoretical fractions complement each other in simulations. This claim will be supported by discussing a further case.

EXAMPLE: A HURRICANE

After treating the subject on a global and even a galactic register, I shall complement the series of examples with fluid flows and spirals on a local scale. "Opal" was the name of a severe hurricane that developed in the fall of 1995 over the Gulf of Mexico and hit the Florida panhandle on the US mainland. This storm caused 59 deaths and property damage to the tune of several billions of dollars. This kind of catastrophe is not atypical for the Gulf region. Policymakers and insurance companies agree on the threat such hurricanes pose.

Hence, there is strong interest in obtaining predictions on their trajectories early enough to take countermeasures. Such predictions have become a significant part of meteorology. This was the context for a research project carried out at the US National Center for Supercomputing Applications (NCSA) and at the University of Illinois Department of Atmospheric Sciences (see Romine 2002, for more details on this project). The project provides further material that highlights the function of visualization; in particular, the visually controlled variation of parameters assumes a guiding role here.

The research project analyzed and (re)constructed a model of Opal that would also be applicable for future hurricanes. A federally funded database stores the trajectory, speed, amount of rain, and further statistics on storm events. The goal of the project was to simulate Opal so that it would fit the known historical data. In a second step, the researchers hoped, it would be feasible to gain predictions by adapting the initial conditions to fit new cases. This strategy of retrodiction in the service of prediction is not unusual. Simulation models are validated using known cases and are then applied to new cases.

The simulation of Opal built on a general model of atmospheric dynamics developed at the National Center for Atmospheric Research (NCAR) in Boulder, Colorado. This basic model, the MM5, or Mesoscale Model Version 5, is an offspring of the general circulation models discussed in chapter 1 that work with a system of fundamental physical equations. In this sense, a hurricane is a "mesoscale" event. It is part of the official policy of NCAR to make such circulation models available off the shelf for other scientific users. These models should serve as a platform for more specialized projects that may add particular modules to the MM5, for example, to adapt the simulation model to hurricanes.

To achieve this, the Opal group considered various parameterization schemes. These were tested in simulation experiments to check whether the schemes were able to deliver a good fit to the actually observed data: "High-resolution MM5 simulations of Hurricane Opal are used to study the impact of various microphysical and boundary layer parameterizations" (Romine & Wilhelmson, 2001, p. 1).

For this task, the adequacy of the assumptions entered into the parameterizations could not be judged independently from the overall performance, because there are only a few theoretical considerations about parameterization schemes.[16] These schemes are devices constructed

16. Sundberg (2007) presents this point from the perspective of the sociology of science.

to achieve a certain performance in an effective way. For instance, the complicated microstructure and dynamics of a cloud is modeled—parameterized—by a scheme at the grid points. Consequently, the adequacy of a parameterization scheme is assessed mainly via the performance in the overall modeling framework. More precisely, the overall dynamics are created only on the basis of concrete parameter values. Thus, the criterion for parameterization schemes was the degree of fit with the historical data that could be achieved through the adaptation of parameter values.

Obviously, the ability to vary simulation experiments in a controlled way played a crucial role. But physical theory was also involved on several levels. One was the fundamental equations of MM5 that went into the Opal project as a plug-and-play model; that is, the theoretical assumptions of MM5 were treated as givens while modeling the hurricane. Also, theoretical assumptions were important on a more heuristic level when constructing parameterization schemes. The latter function within the framework of the larger simulation model and, consequently, their adequacy concerns the entire model and not one of the isolated phenomena or subdynamics that are being parameterized. In this way, different approaches were first implemented on a trial basis and then investigated in an exploratory manner to determine how good a fit they could achieve when parameter values were adapted and refined.

One instance was the parameterizations of the boundary layer—the quote given earlier speaks of "various boundary layer parameterizations." The researchers had a variety of approaches at hand—each theoretically plausible but less than compelling. Should the dynamics be modeled on the micro level or are coarser phenomenological parameters the better and more efficient choice? The resulting overall model dynamics were the leading criterion, but they could not be captured by theoretical reasoning. Thus visualization took on an indispensable role.

The calibration of parameter values relies on the performance observed in simulation experiments. Figure 2.4 displays an intermediate state in which the researchers handle two different parameterization schemes and try to calibrate them so that the trajectory of the simulated hurricane fits the observed one.

This kind of procedure calls for rapid iterations of the loop connecting (simulation) experiment, observation, and new adaptation. Visualization

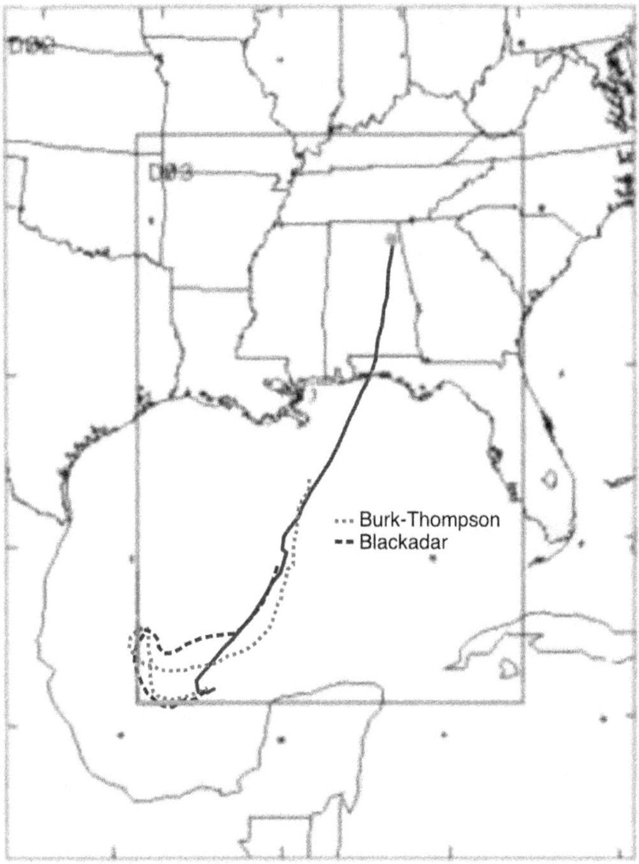

Figure 2.4 Gulf of Mexico and southern US. Solid line: Observed hurricane track to Florida Panhandle and Alabama. Dashed lines: Simulated tracks resulting from two different parameterizations. *Credit: Simulated data as modeled and computed by Romine and Wilhelmson (2001). Courtesy of M. Brandl.*

here works like the "magic eye" of old radio sets that presented the goodness of reception in an intuitive way. The researchers tried to find the right tuning while looking at visual concurrence; the movement of the simulated hurricane should line up with the observed storm track.

As a consequence, the adequacy of the simulation model is not judged by the adequacy of the assumptions that go into the parameterization scheme but, rather, by the results that can be attained. This observation is highlighted by the employment of "black box" schemes: some parameter schemes work with parameters that lack a physical interpretation. It emerged, however, that these schemes were more versatile, so that they could serve better to adapt the simulation model than schemes that work with microphysical assumptions.

Visualizations created the intuitive accessibility of the model behavior that was then utilized crucially in the modeling process. The visual accessibility can help not only to "tune" but also to explore the simulations. After the simulation model has been adapted to the observed data, there is a simulated version of Opal whose trajectory has been optimized. The simulation has much richer internal dynamics than consideration of the hurricane's trajectory covers. Details about the distribution of water vapor, for instance, or about the intensity of vorticity can be visualized separately. In this way, the researchers can study single aspects of the overall dynamics that would be hard to observe *in natura* without artificial coloring or other visualization tools. A series of snapshots, or even a simulated system unfolding on the screen, can exhibit dynamic properties of the model that are otherwise hardly accessible.[17]

AMPLIFICATION

According to the backward logic, the performance attained after the adaptation and calibration process decides whether the modeling assumptions count as adequate. The logic works independently from the origin of

17. In Lenhard (2017b), I discuss dynamic properties in the context of computer-assisted design.

the assumptions—be they either motivated theoretically or of an instrumental nature. Especially in relatively complicated cases such as the ones investigated here, theory can also play an important heuristic role. It seems plausible that this should be the way things work: to achieve a reasonable fit *without* a *theoretical* contribution would be asking too much from modeling, whereas to *exclude instrumental* assumptions and artificial components would presuppose a comprehensive theoretical grip that hardly exists in complex situations.

The indeterminacy and opacity of the assumptions and, in particular, of their interactions that influence the model dynamics appear to be epistemic weaknesses. However, the methodology described turns this into a potential advantage. The point is that a substantial range of modeling assumptions can remain vague and can be determined afterwards—as long as one has ways and means at hand to determine them later in a stepwise procedure that is able to reveal their performative advantages. In this process, visualization assumes a role that amplifies—intensifies and expands—exploratory modeling. Thus the claim that visualization is an important element in simulation modeling that amplifies the exploratory mode has been supported by my discussion of examples.

I would like to conclude this chapter by considering the significance that visualization has for simulation. This concerns the asymmetry between recognizing patterns and creating patterns. It is important to keep these apart, because although the computer is a versatile creator of patterns, it has persistent problems in recognizing them. Put differently, pattern recognition works with similarities that are often hard to formalize.

Keith Gunderson (1985a), for example, has contributed classical papers on the philosophy of mind and artificial intelligence (AI) that conceive of pattern recognition as the key problem of AI. He takes "simulation" in a very specific, today one might say peculiar, sense—namely, as a strong program of AI. According to this program, AI is concerned not only with generating apparently intelligent behavior but also with the more difficult and restricted task of imitating the way in which human intelligent behavior is generated. Gunderson argues convincingly that the prospects for successful simulations (in the strong program) depend on the solution of the problem of automated pattern recognition.

I have discussed simulation in a broader context, not restricted to AI, but as an instrument in a broad spectrum of sciences.[18] Whereas automated pattern recognition did not play a large role there, this did not impede the spread of simulation methods. On the contrary, visualization is utilized as an element of simulation modeling on the basis of a division of labor. In simulation modeling, computers mainly play their role of creating and varying patterns, whereas researchers take advantage of the human capability to recognize patterns.

Based on this division of labor, simulation ventures into complex modeling tasks. Visualizations present a genuine novelty for mathematical modeling—not because human visual capabilities are to be replaced by the computer but, rather, because these capabilities can be incorporated into the process of modeling. The problem of pattern recognition in AI thereby remains largely untouched.[19] All examples discussed here—atmospheric circulation, formation of galaxies, and trajectories of hurricanes—were successful at crucial positions on patterns, but none of the cases required an exact definition and classification of these patterns. Of course, this makes the establishment of a fit between patterns a question of intuitive persuasion. If one could formalize the establishment of a fit, visual output would be rendered unnecessary, or would serve only to display the results that have been calculated beforehand.

These considerations tie in with the findings reported by Peter Galison and others that computer use does not lead to a sort of objectivism based on automation but, rather, that a scientist's educated judgment is required when working with visualizations (Jones & Galison, 1998). Heintz and Huber hold a similar view: "The rehabilitation of human judgment stands in a close relationship with the invention of the computer and the debate it triggered about the differences between machine and human intelligence" (2001, p. 21; my translation).

This point of view is elaborated in Daston and Galison's (2007) volume on the objectivity of scientific representations. I would like to maintain,

18. This spectrum could hardly have been anticipated at the time when Gunderson was writing—that is, in the early 1970s. Today, his usage of simulation would count as outdated.

19. The next chapter will touch upon recent advances in automated classification made by "deep learning" approaches (neural networks).

however, that human judgment does not stand in opposition to computer methods but, rather, it has a role to play in them. It is a peculiar fact that ongoing computerization strengthens the aspect of human judgment in scientific practice. This peculiar and maybe even astonishing observation will be taken up again in the concluding chapter of this book. There, I shall try to balance this observation with the seemingly contradicting—but also justified—verdict of Paul Humphreys when considering the upshot of his studies of computer simulation—namely, that human beings are being driven out of the center of epistemology.

3

Plasticity

Is simulation characterized mainly by the instrumental possibilities it offers to adjust and control mathematical models? That would be in line with Humphreys's (1991) diagnosis of simulation as enlarging the "realm of mathematical tractability"—even though he ties this finding to a consideration of mathematical structure. I shall come back to this claim at the end of the chapter. Indeed, the results of chapters 1 and 2 have shown how simulation experiments and visualizations are employed in the framework of a modeling process—or, rather, would have to be employed if the range of application for mathematical models is to be extended. The present chapter adds plasticity as a further important feature of simulation models. This property contributes to the exploratory and iterative mode of simulation modeling from the side of the model's structure.

A model has the property of plasticity when its dynamic behavior can vary over a broad spectrum while its structure remains unchanged. In all likelihood, there are no models whose behavior is determined completely by their structure. Filling gaps and adjusting parameters have always been part and parcel of modeling. I shall show, however, that simulations present a special case, because plasticity does not appear as a shortcoming that needs to be compensated but, rather, as a systematic pillar of modeling. The structure of the model is designed to include a measure of plasticity that creates room to maneuver. To a significant degree, the performance of the model is then determined during the process of adjustment.

To this very extent, the structure of the model does not aim to represent the (or some idealized) structure of the phenomena to be modeled. Based on the model structure, further specifications are made during later steps of model building, and it is only then that the behavior of the simulation model is determined. In such a procedure, the steps involving the specifications (or adaptations) of the model structure—though normally regarded as pragmatic measures of little significance—acquire a major role in modeling. Thus, the essential claim of the present chapter is as follows: In an important sense, the simulation model does not aim to reproduce the structure of the target phenomena. Instead, it focuses on the opportunities to adapt the model behavior. These opportunities are neither fully independent of nor fully dependent on theoretical structure. As a result, the explorative mode of modeling advances to a status on a par with theory-based considerations about structures. The relationship between the two is not fixed strictly, but can be balanced in various ratios.

Because plasticity and the accompanying necessity for a phase of specifications come in various degrees, it can play a minor role in some cases but a major role in others.

I would like to distinguish two different aspects in models: (a) their structure and (b) their specification. This distinction is relevant because the structural part of models is incomplete in an important sense: the model behavior is not derived from the structure.

There are still significant gaps—even gaps of a conceptual kind. A given structure can give rise to quite different dynamics depending on how variables, parameters, or entire modules are specified. Hence, the model dynamics can be called *structurally underdetermined*. Only the process of specification—which involves the activities of experimenting and visualizing as I have shown here—fully determines the model behavior. In short, plasticity and exploratory modeling complement one another. Because of a model's plasticity, the exploration of model behavior addresses more than minor pragmatic issues of adaptation; however, the performance of models with a high plasticity can be investigated and controlled only in an exploratory mode.

PLASTICITY AND SPECIFICATION

"Plasticity" is a common term in biology, in which it means the ability to adapt. Stem cells, for instance, are of particular interest for medicine because they can develop a nearly universal functionality depending on which cell compound they belong to. Hence, these cells represent ideal cases of plasticity. The structure of the cell does not predetermine its further specification. Instead, this is done only when the cell develops in a specific context. In this way, the structure is decisive insofar as it enables or encompasses all these possibilities. Simulation models are comparable.

Here, too, plasticity is a matter of degree. Admittedly, it is hard to find any model dynamics that would be determined completely by the model structure; hence, there is always at least a small degree of plasticity. However, structure plays a leading role in traditional exemplars of mathematical models. The mathematical structure of Newton's theory of gravitation, for instance, already determines that the gravitational force varies with the quadratic distance, that planets move on elliptical orbits, and that the actual strength of gravitation on our planet depends on an empirical constant. I would like to emphasize that mathematical structure is important, but it is not everything. This holds for any mathematical modeling—whether simulation or not. The thesis of structural underdetermination intends to signal more—namely, that *simulation* models show a particular *plasticity*. Then, the structure is not sufficient to determine model behavior even qualitatively. Plasticity and structural underdetermination are two sides of the same coin.

This claim may appear contraintuitive, and I would like to recall that this is addressing degrees, and that different simulation models might be structurally underdetermined to different degrees. As I shall try to argue and substantiate through examples, the remaining gaps can be closed by conceptual means, as well as by techniques of adaptation. It is exactly the particular requirement to close this gap during the modeling phase that makes the characteristics of simulation modeling encountered here so indispensable—namely, that it proceeds in a stepwise and exploratory backwards logic.

Simulation models, considered only according to their structure, are model schemata that require further specification. The modeling process therefore has to include an activity that experiments with this schema and that specifies and calibrates the model behavior. The last point does not just refer to those parameters whose values are assigned by initial and boundary conditions (as in Newton's theory). It can also deal with fundamental aspects of the model's dynamics, as some examples will show. So let me take stock: the differentiation of modeling into structure (schema) and specification depends on the plasticity of models; that is, it depends on the claim that when the structure is kept fixed, the process of specification can influence and change the dynamic behavior in fundamental ways.

If this claim is correct, it challenges longstanding conceptions of what mathematical models do. After all, structure counts as the core of a mathematized scientific theory. Such a theory is not taken to represent phenomena directly; rather, the philosophical discussion views it as a reduced or idealized description that is targeted on the structure of phenomena. To describe this structure, mathematical language is seen as an adequate, or even the only adequate, language. Therefore, the thesis of plasticity and the accompanying division into structure and specification imply a significant shift in the conception of mathematical modeling. Simulation is oriented toward performance, as has been discussed in chapters 1 and 2. This is not a mere instrumental bonus of simulation compared to traditional mathematical models but also affects its conceptual core. To the extent that the structure does not determine behavior, it also falls short of representing the essential dynamic characteristics.

This observation is directly opposed to the common opinion that mathematical models bring out the essential structure by representing the real relationships in an idealized way and leaving out what is negligible.

I would like to point out that the particular question regarding how simulation models should be characterized as a type of mathematical models is of a broader philosophical relevance. An epistemological core area of modern science is the way in which knowledge is gained and organized in a mathematical form. What is meant by mathematization, in turn, is closely related to the conception of model and structure. In his *Substanzbegriff und Funktionsbegriff* (*Substance and Function*), Ernst Cassirer (1910) argued that the modern standpoint is characterized by a transformation that replaces the Aristotelean primacy of objects with a relational perspective that ascribes fundamental significance not to the objects but to the relationships between them. Thus, the concept of function becomes fundamental and replaces the ontology of substances with an ontology of relational structures. This philosophical approach is based on the success of mathematical modeling in the natural sciences.

In recent discussions, what has been termed "structural realism" holds a somewhat related position. It admits that there are many ways in which scientific knowledge does not deal with how things really are. In one sense, however, a realist interpretation holds that the part of scientific knowledge that deals with reality is the one about relational structures. James Ladyman (1998), for instance, is a prominent proponent of this, even though his view differs greatly from Cassirer's. This underlines the broad relevance: functional thinking and structural realism in its various forms (see also Worrall, 1989) share the belief that the essential core of scientific knowledge lies in relational structures expressed by mathematical functions.

However, simulation models are types of mathematical models in which this reasoning runs into trouble. This happens because the strength of mathematical modeling in simulation models does not coincide with the strength of mathematical structure. Simulations work despite the models being structurally underdetermined,[1] or even because they are (if my claim is right). The success of mathematical modeling cannot then count as an argument for realism. Simulation presents a shift in the conception of mathematical modeling. Since this shift, the basis of the argument about relational thinking appears in a new light. I do not intend to enter into the debate on structural realism here. The concluding chapter 8 will discuss how simulation challenges the entrenched philosophical divide between realist and instrumentalist positions.

Let me return to the main flow of my argument. In the following, I shall analyze important types of simulation and discuss cases supporting the thesis that simulation models are structurally underdetermined and that underdetermination comes in degrees. The first topic will be artificial neural networks (ANNs). These show a very general behavior; that is, their behavior can vary broadly, although these models are based on a practically generic architecture of (artificial) neurons and synapses. The input/output patterns produced by such a model depend almost completely on the weighting of the connections in the network—that is, on

1. A main motivation of structural realism is to defend a realist position that acknowledges *empirical* underdetermination.

the values assigned to the parameters. A second type is finite differences in which the models are often founded on a rich theoretical structure. Here, I shall return to the investigation of fluid dynamics and meteorology. Specification does play a significantly smaller role here than in ANNs, but a certain degree of structural underdetermination remains. As my third and last class of examples, I shall discuss cellular automata (CA). In terms of the degree of their structural underdetermination, they can be located in between the first two types.

ARTIFICIAL NEURAL NETWORKS

Just like most types of simulation, artificial neural networks (ANNs) go back to the pioneering age of computer development. Since the groundbreaking work of McCulloch and Pitts (1943), ANNs have developed into a prominent type of simulation model that has now grown into a whole family. For the present concern, a rough picture suffices. Basically, ANNs comprise a number of connected neurons (knots in the network)—artificial, idealized counterparts to physiological neurons. Each artificial neuron contains a soma with a number of incoming and outgoing axa including weighting factors. The neurons work according to a simple rule: all incoming signals are added up according to their weighting factors, and as soon as the sum crosses a certain threshold, the neuron "fires." That means, it sends a signal that is distributed along the outgoing axa according to the weighting factors. In a network, the neurons are placed in layers so that a neuron of layer x has outgoing connections to neurons of layer $x + 1$ (in a noncircular network). Figure 3.1 displays a small ANN in which the weighting of the connections is symbolized by the thickness of the arrows.

ANNs can have one or several intermediate layers. As soon as the connections and weighting factors have been determined, an ANN transforms an input pattern—that is, an assignment of initial values to the first layer—into an output pattern, which is the resulting values of the last layer. Altogether, this sounds like rather simple dynamics. The point is the following: By modifying the weighting factors, the input–output dynamics can be changed in unexpectedly variable ways. The early and

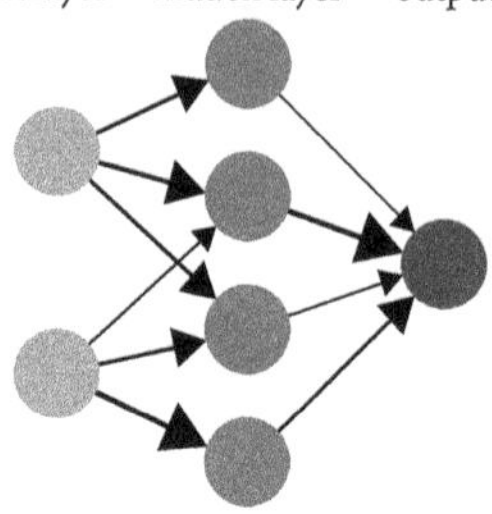

Figure 3.1 A neural network with one intermediate layer. *Credit: Wikimedia Commons.*

famous discovery of McCulloch and Pitts about ANNs was that a rather rudimentary class of ANNs is sufficient to produce a wide range of dynamic behavior—namely, all Turing-computable patterns. This seemed to elevate ANNs to a universal model schema and fueled the early *computational view* on intelligence and the human mind, because one could now point to a seemingly direct correspondence between brain and computer. ANN architectures were conceived as models, if highly simplified ones, of the mechanisms at work in the brain. Hence, ANNs attracted special interest as an object of investigation: their architecture was taken from the physiology of the brain while their dynamics were easily accessible for mathematical or simulation-based modeling.

However, it turned out that the actual use of ANNs posed many problems, because it is hard, in principle, to transform the universality of the models into something concrete that works in practice. The models had a degree of plasticity that was too high, and there was a lack of methods that would help to assign a specification that would be both adequate (regarding the patterns) and efficient (regarding the required time for calibration). Therefore, ANNs remained somewhat distanced from practical applications, and this was arguably a main reason why they went out of fashion toward the end of the 1960s. Since then, sophisticated learning algorithms have led to a renaissance. They are based mainly on what is called *backward propagation*, a stepwise strategy to adapt the weighting factors so that the desired input–output behavior is approximated. One speaks of the networks "learning" the appropriate behavior. During the course of the learning process, the structure of a network remains practically invariant. The enormous variability in possible behavior is attained by the specification—that is, the assignment of values for the weighting factors and thresholds.

A recent example is the commercial software package Neurosolutions 5.0 that is advertised to help with "machine diagnostics, speech recognition, portfolio management," and some further tasks (Neurosolutions, 2007). Whether the software really lives up to its promise for all the different tasks remains an open question here. What is more important is that the idea behind marketing the product works with the plasticity of the model. The company sells one and the same network as a kind of platform on which to work toward the solution of quite different problems, as well as a method of specification to adapt the platform to different applied contexts.[2]

Accordingly, an ANN in which weighting factors are not yet assigned presents more of a schema of a model rather than a model proper. Such an ANN does not give a mapping from input to output patterns; instead, it gives an entire class of possible mappings. One such mapping is singled out only when the exact weighting has been assigned. The point of ANNs consists in their large (or ultra-large) plasticity. They can serve as a platform[3] exactly because their dynamics are structurally underdetermined and hence allow high adaptability.

Plasticity, however, comes with a downside: ANNs had attracted the attention of modelers because they appeared to be so similar to real physiological networks. But this structural similarity now loses its significance, because a structural similarity has little relevance when it can give rise to a quasi-universal dynamics. Moreover, the weighting factors are assigned in a process that optimizes overall performance, and nobody would claim that the resulting weights had a physiological counterpart. Hence, a

2. The software package SWARM developed at the Santa Fe Institute presents an example of a platform comprising a design concept and a program library.
3. The notion of a "platform" has a connotation linking it to mass production and technological economy, as when two types of cars are manufactured using the same platform. Whereas this connotation fits well when discussing the example of the Neurosolutions software, the fit is not so good with other examples in which plasticity is not related to economy.

successful application of an ANN—say, in speech recognition—is of limited epistemic value when the aim is to investigate how speech recognition works in the human brain.

The strategy termed "Deep learning" is heralded as a key way of extracting and utilizing information from big data. These strategies work with ANNs that include hidden layers of neurons, which is why they are called "deep." Learning algorithms have made progress and, together with computational power, this makes training an ANN a much more feasible task than it was a couple of decades ago. In my eyes, one major philosophical problem with deep learning is epistemological. Deep-learning algorithms can achieve remarkable feats while human beings struggle to learn something from the model beyond the mere fact of the trained input–output relation.

I do not want to suggest that modeling with ANNs would be mere tinkering. There are important theoretical insights at play. One example is the adaptation behavior of certain network architectures that refers, however, mainly to the mathematical properties of the artificial networks themselves.[4] One thing seems to follow: success in simulating a certain phenomenon does not justify concluding that the structure of this ANN model would in any way represent the structure of the phenomenon being modeled. Instead, a sort of implicit knowledge is at play: "In a fairly intuitive sense, it seemed correct to say of a trained-up network that it could embody knowledge about a domain without explicitly representing the knowledge" (Clark & Toribio, 1994, p. 403).

Therefore, artificial neural networks exemplify in an outstanding way the thesis that simulation models are structurally underdetermined and that one can distinguish schema (platform) from specification. It has to be admitted that this is an extreme case of plasticity in the field of simulations. Plasticity comes in degrees, as I have mentioned. The next case comes from the opposite end of the spectrum where models with significant theoretical structure emerge in which the thesis of plasticity initially appears to be much less likely.

4. Elman (1994), for instance, relies on plasticity to show how an ANN with "short-term memory" gets better at solving certain problems when the short-term memory is enlarged.

FINITE DIFFERENCES

The method of finite differences, together with its cousin the finite element method (FEM), are of particular importance as they are employed ubiquitously in the natural sciences and engineering. This is because they start from mathematical descriptions that are particularly widespread in these areas.[5] The finite difference method (FDM) was introduced in chapter 1 as a way of treating systems of differential equations numerically. FEM follows a related approach. It splits spatial components into smaller (but finite) elements, considers the behavior of each element in relationship to its neighboring elements, and in this way reassembles the global dynamics.

In a sense, both methods take one step back from integration. This can be derived from looking at the limit processes in which an increasing number of smaller and smaller elements finally give way to an integral. Both simulation methods work in opposite directions and replace the operations of analytic calculus through a great number of single steps. The key point regarding the range of applicability for these numerical methods is that they can disregard the limits of analytical tractability. At the same time, the enormous number of computational steps that result are delegated to the machine.

In a similar vein as in chapter 1, these finite methods present the "hard" case for the claim about plasticity and structural underdetermination. This is because such discretized simulation models rely on continuous models that, in turn, are based on theoretical accounts. Hence, it seems to be justified to consider that discretization—via FDM or FEM—would mainly operationalize the theoretical structure of the continuous counterpart. Then, the continuous models could be claimed rightly to be the structural basis for the simulation. This point is largely correct. The plasticity is substantially smaller compared to artificial neural networks. Plasticity comes in grades, and these grades are not only case-dependent but also relate

5. Johnson (2004) contributes one of the surprisingly few studies on this topic in philosophy and history of science.

strongly to types of simulation techniques. In general, finite elements and finite differences show a relatively low degree of plasticity.

However, this does not mean that such simulation models are determined completely by the structure of their continuous counterparts. There are several reasons why this is not the case, and I have already mentioned some of them. First and foremost, models in general are not completely determined from the side of theory. Regarding this point, and notwithstanding other important differences, analyses in the spectrum spanning Cartwright and Morrison (cf. Cartwright, 1983; or the edited volume by Morgan & Morrison, 1999) coincide. Winsberg (2003) has convincingly exemplified the partial autonomy of FDM simulations. I agree completely on this point. The step from the theoretical and continuous model to the discrete simulation model is particularly relevant.

Generally, there are several options regarding how to discretize. Even if a continuous theoretical mathematical model serves as the starting point, the properties of the related discrete model might turn out very differently, because the various options for discretization do not coincide when it comes to the dynamic properties of the simulation model. The lack of coincidence normally affects the simulation models when compared among themselves, as well as compared to the basic theoretical model. Hence, one has to accept discretization as a fundamental step in modeling, a step that is specific to simulation. Therefore, the considerations about plasticity of modeling apply repeatedly.[6]

When one has decided to work with a specific discretization, adapting the dynamic behavior of the simulation to the target phenomena requires further measures. These include, in particular, the employment of nonrepresentative artificial components (Lenhard, 2007; Winsberg, 2003), as discussed in chapter 1 for the case of the Arakawa operator. Such components constitute provisions for dealing with and specifying the extra plasticity that has been introduced through the discretization.

In other words, discretization produces a gap between the theoretical and a continuous mathematical description of the phenomena and their simulation. And, in principle, this gap cannot be closed by referring to the

6. Küppers and Lenhard (2005) describe simulation as modeling of a second order.

theoretical structure. Expressed in this way, it is not so much remarkable as more to be expected that the process of simulation modeling will have to include compensatory elements that counteract the unwanted effects of discretization.

One important way of making room for plasticity is to employ parameterization schemes. I mentioned such schemes in the previous chapter, and I shall discuss them again later in the book and also in this chapter. They are usually based on theoretical knowledge about which kind of parameters and which sort of interdependency might make sense. At the same time, parameterization schemes insert plasticity because the parameter values influence the model behavior. In this way, they can compensate for discretization errors even without the researchers knowing them explicitly. Hasse and Lenhard (2017) support this point in their study of simulation modeling in thermodynamics engineering. Equations of state are based on sound theory, whereas at the same time they include a host of adjustable parameters that allow the application of the equations in practical (nonideal) cases.

The case of atmospheric circulation discussed in chapter 1 illustrates the relationship between plasticity and the guiding role of specification. Hence, there are two effects that run counter to each other and are balanced out in simulation—at least in successful instances. On the one hand, simulation creates structural underdetermination; that is, the gap between the theoretical structure and behavior of the model broadens. On the other hand, simulation offers a remedy; that is, an instrument to bridge this gap: the employment of an iterative and exploratory mode of modeling might balance structure with specification in relation to the overall model dynamics.

Notwithstanding the significance of plasticity, it should be clear that cases such as finite differences or finite elements differ considerably from artificial neural networks. In the latter, the structure could function like a platform that could be specified—that is, adapted to *very different* requirements—whereas in the former case of the finite difference method, the process of specification served to simulate one *particular* complex dynamic. As a general rule, a discrete model works more like a bridgehead than a platform—a point from which the specification can start to move toward a certain goal.

CELLULAR AUTOMATA

A couple of authors consider cellular automata (CA) to be the one class of models that brings out the essential characteristics of simulation. Be it that there is a new syntax at work (Rohrlich, 1991), a new kind of modeling "from above" (Keller, 2003), or even *A New Kind of Science*, as Stephen Wolfram's lengthy book has it (2002; see also Barberousse, Franceschelli, & Imbert, 2007, who discuss a sample of CA methods). All the authors mentioned here, by the way, underline the role played by visualizations. This is grist for the mill of the plasticity thesis, because interacting with visualizations offers manifold possibilities for changing and adapting CA dynamics in a nuanced way. Dynamic visualizations provide an effective means for controlled specification.

In general, simulations employ a discrete generative mechanism for producing—or reproducing—the dynamics of a process. CA rest on a surprisingly simple class of such mechanisms, as discussed briefly in chapter 2. Quite simple local rules govern the interaction of cells and are able to generate very complex global dynamical patterns. One advantage of the CA model class is that the computer can apply its particular strength—namely, to make myriad serial local adaptations. A theoretical model may serve as guide for the CA model, as was the case with the development of spiral galaxies discussed in chapter 2. CA mechanisms might also be found as generators in their own right. An example is John Conway's "Game of Life" that became famous in the 1970s and was popularized by Martin Gardner (1970).

Conway had been experimenting with various simplifications of Ulam's approach to the theory of automata, with the goal of generating interesting kinds of patterns. In other words, he wanted to generate dynamically stationary patterns. One example of such a pattern is the glider gun displayed in figure 3.2. In every step, each cell is occupied (or not) according to a particular rule that depends only on the states (occupied or empty) of its neighboring cells. Following this rule, the pattern in the upper third of figure 3.2 changes in a stepwise manner. After a couple of steps, the pattern returns to the initial form, but has generated an additional small pattern of five occupied cells. Following the very same rule according to which cells are occupied and emptied, the small pattern moves to the lower right of figure 3.2. This creates a series of these small patterns that seem to move through the grid space—like gliders. The regularly growing number of gliders that move on independently can then count as proof of the existence of patterns that will grow to infinity in a controlled way—that is, without occupying the entire grid.

However, theoretically based models represent the greater challenge for the thesis of structural underdetermination, and this is why they are receiving the major part of my attention in this section.

Seiden and Schulman's CA model of the development and formation of spiral galaxies (see chapter 2) was discussed already by Rohrlich (1991). I would like merely to add some details in order to clearly reveal the sense in which CA is in line with other simulation modeling approaches when it comes to plasticity. Seiden and Schulman utilized a percolation model to test whether the theory of stochastic self-propagating star formation

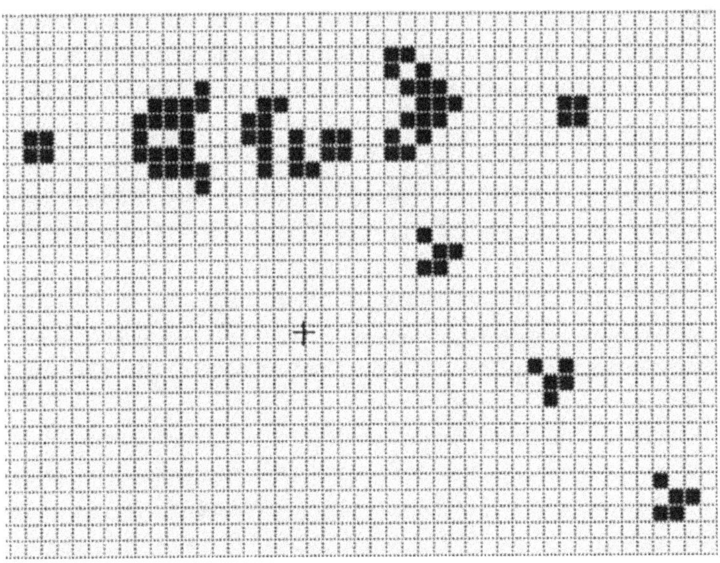

Figure 3.2 Snapshot of a glider gun: Small clusters of five cells glide to the lower right while the configuration in the upper part periodically generates new gliders. *Credit: Wikimedia Commons.*

(SSPSF) could provide a plausible picture of the characteristic spiral forms of observed galaxies. Their "dartboard" (see figure 2.2) served as the initial state of a cellular automaton, and the dynamics between cells should model the essential aspects of physical processing going on in such conglomerations of stars. The success criterion was an intuitively very graspable one: the visual fit between simulated and observed patterns. However, Seiden and Schulman actually had a rather theoretical aim—namely, to show that the stochastic theory (SSPSF) was adequate and productive. To reach or tackle this aim, they had to resort to simulations, because neither the internal dynamics of SSPSF nor the comparison with observations would have been accessible by any other means.

In this case, it is not obvious what exactly should count as the structural part of the CA model. The annular shape of the dartboard clearly results from theoretical considerations that suggest such a kind of structure. On the other hand, how the velocity of rotation varies on certain rings depends on the radius and duration of the movement. This variation belongs to the specification part. Or rather, the need for *some* parameter that guides the dependence and variation is a structural concern. Then, adapting this parameter is a matter of specification, and this can take place only in an exploratory mode. Furthermore, it was not theoretically necessary for any visual agreement to be attained at all for a particular parameter configuration. Unlike the case of the gravitational constant whose existence is given by theory, in the present case it was an open question that could be addressed only in an exploratory mode.

Hence, important modeling assumptions are founded on theoretical considerations—at the end of the day, the results of the simulation should be significant for SSPSF—and the interaction between the cells should represent physical processes such as dust explosions.

Such structural assumptions, however, do not determine the observable model dynamics. These have to be supplemented in an important way. Form and structure of the visualized results depend on the structure, but significantly also on the exact specification of the dynamics. For instance, theory does not stipulate how the rotation of the disc should be controlled or how the partition of cells should be refined during runtime. These are questions that concern the nature of the CA model, and they can be answered only by observing the model dynamics in a quasi-empirical way. The properties of the CA model, in turn, are not related to physical theories in any transparent way. To give an example similar to the case of finite differences, there is no concept of distance according to which one could distinguish "fast" from "slow." Partition and rotation follow a dynamic that is internal to the simulation model and is controlled merely by the observed—that is, the visualized, global performance.

Therefore, claiming that theoretically motivated decisions about parameterizations are tested visually through simulations would oversimplify things. It is far more the case that the adaptation to data plays a crucial role in the modeling process itself. Both the use of visualizations and the extensive reliance on exploratory simulation experiments during the process of model specification are typical for CAs. If their various attempts to adapt the CA model were to have failed, Seiden and Schulman would have had to resort to other parametrization schemes. Probably they did in fact undertake such trials, but as we all know, mainstream scientific publications hardly ever include unsuccessful attempts.

Again, as in the case of finite differences, the structural assumptions about the CA model are based on a theoretical model or a theory. At the same time, however, important conceptual gaps can be closed only during the process of specification. In short, what appears to be a shortcoming when compared to the standard of mathematical modeling turns out to be an advantage in the context of complex systems.

A second example will confirm this finding. The lattice gas automata once again originate in the field of fluid dynamics—this time seen from the perspective of CA modeling. Brosl Hasslacher, who worked at the Los Alamos National Laboratory, followed an approach based on CA to study complex problems in fluid dynamics. In his report *Discrete Fluids* (Hasslacher, 1987), he summarizes that "until recently there was no example of a cellular automaton that simulated a large physical system, even in a rough, qualitative way" (p. 177). Hasslacher and his coworkers undertook to create such an example.

They wanted to construct their CA model so that it would correspond to kinetic theory; that is, the dynamics resulting from the interaction of cells should correspond to the dynamics expressed in the Navier–Stokes equations. Accordingly, their criterion for success was whether they could reproduce the results already known and accepted from the theoretical equations. As in the first example, the test required a simulation: "that one derives the Navier–Stokes equations with the correct coefficients and not some other macrodynamics—is justified after the fact" (Hasslacher 1987, p. 181). What Hasslacher circumscribed here as "after the fact" does not just mean that simulations are necessary to bring to light the implications of the modeling assumptions. It goes beyond this and also means that additional specifications are indispensable in order to amend the structural assumptions. Only the successful specification—and not merely the structural core—was able to justify the entire modeling approach.

Hasslacher started with a couple of essential structural assumptions; most notably, his choice of grid was motivated by theoretical considerations. To be specific, the cells were arranged in a hexagonal grid for building the "Hexagonal Lattice Gas Automaton," because it can be shown that for reasons of isotropy, no rectangular grid can give rise to the homogeneity properties shown by the Navier–Stokes equations. The hexagonal grid is the simplest arrangement that suffices to meet these theoretical requirements for isotropy.

However, this underlying geometrical structure of the model in no way determines particular behavior. Extensive simulation experiments are already necessary to gain a detailed dynamic behavior of the model, because there is no available theoretical guideline that would prescribe how to calibrate the parameters of the cellular automaton that control the local interactions between cells. Thus the researchers could only hope to detect agreement between simulated and theoretically known behavior in an iterative–exploratory way. Such agreement, then, could be accepted as sufficient motivation for validating the (after-the-fact) successful assumptions. Just as in the cases discussed earlier, the pivotal criterion was the performance of the simulation model—in particular, whether it would behave in a similar way to Navier–Stokes in a global respect: "In general, whenever the automaton is run in the Navier–Stokes range, it produces the expected global topological behavior and correct functional forms for various fluid dynamical laws" (Hasslacher, 1987, pp. 198–199).

It took a combination of structural and specifying parts to deliver encouraging results for Hasslacher. Nonetheless, the agreement could not guarantee that the simulation would approximate its continuous counterpart, for structural reasons. Indeed, this is not the case, notwithstanding the remarkably good agreement Hasslacher observed. One can see this from the principal differences introduced by the discrete dynamics. In particular, the simple lattice gas automaton has a Fermi–Dirac distribution, whereas the continuous fluid dynamics generate a Maxwell–Boltzmann distribution. Consequently, during the process of specification, the effect of this difference has to be compensated by certain—artificial—extra assumptions. Hasslacher (1987) succeeded in minimizing the influence of this difference on global behavior without overcoming it:

> Simulations with the two-dimensional lattice-gas model hang together rather well as a simulator of Navier–Stokes dynamics. The method is accurate enough to test theoretical turbulent mechanisms at high Reynolds number and as a simulation tool for complex geometries, provided that velocity-dependent effects due to the Fermi nature of the automaton are correctly included. (p. 200)

Briefly stated, the Fermi–Dirac character of the CA simulation could be compensated in certain limits by clever specification. The plasticity of the CA model created space for artificial extra assumptions with the goal of generating the desired dynamics.[7]

PLASTICITY AND STRUCTURAL UNDERDETERMINATION

In the previous sections of this chapter, I discussed various cases and modeling techniques. Let me summarize the results. A series of examples illustrated the process of modeling in different classes of simulations. Two invariant components of the modeling process could be discerned: first, the structural part that leads to a sort of model schema, a class of potential models; second, the specification during which the schema forms a particular model dynamics. In the course of this investigation, different

7. The discussion on CA simulations has included a mix of bottom-up and top-down considerations, so that it would seem to be misleading to identify CA with one of these approaches—even though some opinions expressed in the literature do so.

kinds of schemata have emerged—that is, different kinds of model structures: artificial neural networks with variable weighting parameters; a system of finite-difference equations on a grid, but (still) lacking artificial compensating elements; and cellular automata in which only part of the dynamics, such as the ring-shaped form or the hexagonal geometry of neighborhood, is determined by the structure, whereas other parts of cellular interactions remain open so that they can be specified in an exploratory mode.

Some of these schemata were based quite directly on accepted theoretical foundations such as those of fluid dynamics. In these cases, specification is necessary to compensate for structural deficits or to find a balance by using either parameter calibrations—after-the-fact—or artificial elements. The other end of the spectrum contains schemata that serve as generic platforms such as the ANNs of Neurosolutions in which the steps of specification can even be marketed as advantages in the service of applications. In such cases, the model structure is largely independent of the context of application being considered. That means that hardly anything relevant about the phenomena to be modeled can be inferred from the model structure.

Notwithstanding these differences between types of models, it is clear that the structure, or the schema, requires additional specifications to determine a particular model dynamics. Hence, one can speak of *structural underdetermination* in general, even if it comes in different degrees.

The characteristics of simulation investigated in the previous chapter are relevant mainly for the process of specification. Roughly speaking, only the means for specification that the researchers have at hand—exploratory experiments, visualizations—enable them to start with classes of underdetermined structures. Without these means, the expectation of attaining manageable models would be low. Under such circumstances, plasticity is not a deficit but, rather, an asset that is tailor-made for simulations.

Specification is a process and does not need be carried out in just one step. Instead, the dynamic performance of a simulation can be determined and adapted in a step-wise procedure. Naturally, this process cannot be based on theoretical assumptions, because its main goal is to compensate for the deviations that occur between theory and discrete models. More precisely, this does not mean that the modeling steps are independent from any theory, but that the relevant theoretical insights and assumptions will be about the instrument and the method and not about the phenomenon to be modeled. An example is parameterization schemes that are introduced because of their numerical and algorithmic qualities.

At this point, it is appropriate to discuss a viewpoint developed and advocated by Paul Humphreys that seems to contradict the thesis of plasticity as an important characteristic.

Humphreys (2004, chap. 3.4) underlines the role of what have been termed "computational templates" as essential building blocks for simulations. Such templates are mathematical elements that belong to the structure and have a broad range of applications. Basically, one can compare them to equations or systems of equations that are able to represent the dynamics of not just one particular field but also several diverse fields. For instance, the distribution of temperature in a body and the distribution of particles that move according to Brownian motion have the same mathematical form. In addition, the Laplace equation is relevant for describing not only fluid flow but also electric fields. Hence, once again, it looks as if it is the deeper mathematical structure that matters, and that it is this that has to be made numerically or quantitatively accessible via computer modeling. In short, computer methods make for "tractability," whereas applicability is granted by the deeper structure.

I do not see a contradiction to the thesis about plasticity and underdetermination. Humphreys is formulating a valid point about the flexibility of model structure. Computational templates can be used and reused in different contexts, making model structures generic (to a certain point). In the preceding discussion, I shifted attention to the *additional* processes of specification that are necessary when embedding a structural core into a particular context of application. Whether fluid dynamics (in general) or a model of the atmosphere (in particular) is the issue represents a significant difference. Even if *computational templates* can be identified, various adaptation procedures are necessary going beyond the templates.

THE DILEMMA OF PLASTICITY

Plasticity leads to the following dilemma: To the extent that the model dynamics and the quality of the simulation do not rest on the structure of the model, this structure cannot serve as the source for an explanation of how the simulation works. Moreover, if such an explanation depends on how particular parameters and so forth are balanced against each other, it dissipates entirely. Pointing toward the interplay of these particular assignments would be more like admitting the absence of any explanation.

In other words, including the specification in the picture of simulation can lead to the accusation that one is taking an ad hoc approach; that is, one is following an instrumentalist viewpoint. In contrast, if one excludes specification and concentrates on structure, one has to face the accusation of being arbitrary, because the plasticity of models implies a relative lack of structure.

What the present analysis suggests is in fact the case: this dilemma accompanies simulation. One instance is the criticism of climate models—more precisely, the controversy surrounding the so-called flux adjustment that found its way into the mass media. In the mid-1990s, the circulation models of the atmosphere and the oceans could be coupled successfully. Both circulation models, however, had initially been specified and calibrated separately. The coupling threatened to drive both models out of equilibrium. This was an unwanted effect, because the results should be about systems in equilibrium. A pragmatic proposal to solve this was to compensate the flux between atmosphere and oceans so that both spheres remained in equilibrium.

The criticism was that ad hoc adjustments would be used to fake a realistic appearance, although flux adjustment was in fact a parameter used to control the behavior of the simulation that had no counterpart in reality. The majority of modelers met this criticism with incomprehension, because, for them, balancing artificial parameters is part and parcel of the simulation method.[8] This dilemma comes in degrees, as the following examples will show.

8. Even if adjustments are unavoidable in simulation, this does not mean that arbitrary ad hoc adjustments would be unavoidable as well. Climate modeling relatively quickly devised alternative procedures that worked without the disputed flux adjustment. Regarding this

Each of the following three examples goes back to precomputer times, although all of them came into widespread use only through simulation methods. The first case is the Ising model of percolation theory that starts with magnets arranged on a grid of one or more dimensions. The polarity of a magnet at a certain location is influenced by the polarity of its neighbors on the grid. A parameter of thermic energy controls the degree to which the states at grid points are independent from their neighbors. Thus, the Ising model is a kind of cellular automaton, though Ernst Ising introduced it in 1925 as a contribution to statistical physics. The model is famous because its dynamics include critical phenomena and phase transitions, although these dynamics are defined in very simple terms. Some results, such as the existence of phase transitions or estimations of upper and lower bounds for critical parameters, could be obtained analytically, in particular, by stochastic methods. At the same time, these models were notorious for their computational complexity. Often it was completely impossible to quantitatively determine the interesting parameter values.

Owing to simulation, the situation has changed completely. Markov chain Monte Carlo (MCMC) methods turn challenges such as determining critical parameter values in the Ising model into common exercises in courses for stochastics, statistical physics, or computational physics (the disciplines overlap). Simulation experiments and visualization can help to investigate critical phases efficiently. Figure 3.3 displays an example of a cluster that has been generated to examine the geometrical properties of critical clusters that have the same polarity. Hence, it exemplifies the exploratory mode of simulation.[9]

The second example is the closely related "Game of Life" mentioned earlier. At the beginning, Conway had experimented with coins, buttons, and the like to find a promising dynamic or, rather, a promising class of dynamics. Soon after he succeeded and had defined the game, great enthusiasm set in and all sorts of variants were tried on computers. In this way, Conway's prize question was answered very soon (he had offered

case, see the *Third Assessment Report* of the IPCC (Houghton et al., 2001) that expounds a mediating position.

9. Hughes (1999) gives an excellent and more detailed discussion of the Ising model.

Figure 3.3 A cluster of an Ising model near the critical temperature.
Credit: Wikimedia Commons.

a reward of $50)—namely, whether a controlled and infinite growth is possible. The glider gun (figure 3.2) gives such an example. The answer to Conway's question can be considered as an answer to the question whether self-replicating automata exist—a question that had initially motivated Ulam and von Neumann to think about cellular automata as a model class.

Finally, the third example is social simulations. Thomas Schelling conducted his famous simulation experiments in order to study social segregation. He utilized a two-dimensional grid, like one for cellular automata. However, instead of magnets or cells changing their state, inhabitants show a tendency to change the location in which they live. Their tendency to move is related to the properties of their direct neighborhood (such as skin color).

Schelling also started to work with hand-drawn maps and improvised means.[10] However, the method only became widely accepted with the advent of the computer.

10. See Knobloch (2009) for a detailed discussion.

All three examples have one thing in common: each achieved great success by treating a question in a way that had been out of reach before. What does their explanatory power consist of? Primarily it rests on simplicity and efficiency. Phenomena of phase transition can be studied in the Ising model; hence, more complicated physical assumptions can be left aside. The same applies to the "Game of Life." Automata theory becomes an experimental game that can be restricted to a very narrow modeling framework, because the simple case is rich enough. And Schelling's model for segregation showed convincingly how little of social theory has to go into a model able to generate segregation.

These results, however, do not speak for simulation as a science of structure, because they rest on the full specifications and would not be possible without them. Several authors speak of "how possibly" accounts, or stress the "what if" character of varying specifications (see, e.g., Grüne-Yanoff, 2009; Ylikoski, 2014). Single cases may appear spectacular: "Look! Social segregation can emerge that easily." But how general are the insights such cases can produce? How closely is the simulation related to social segregation in particular locations and societies? According to the logic of argumentation, generating the simulated patterns of segregationist behavior does not require more complicated model assumptions. Whether and how one could build a simulation-based science based on findings of this kind is controversial. One direction picks up the aspect of playing around with the models: The family of models is rich enough to allow a great range of variants; consequently, the issue then is artificial societies with increasingly complex interactions. The fields of "artificial life" and of social simulations build a kind of subdiscipline[11] that follows this path. Another example is the simulation-based analysis of certain game-theoretical situations such as the famous prisoner's dilemma. In short, fields such as these take their simulation-based approaches as objects of investigation in their own right that are constituted and analyzed by increasingly sophisticated and complex models.

A second line is more concerned with the significance of simulation results going beyond aspects internal to the model. After determining that the phenomenon of segregation was a possibility in his model, Schelling

11. On social simulation, see Epstein and Axtell (1996). See Boden (1996) for an overview.

had already directed his attention toward investigating how robust the phenomenon was when the parameters were varied. The goal was to plausibly transfer the simulation to the social world. Here, we face the problem of plasticity in all its strength: Because one can implement very different dynamics on neighborhood grids of the sort Schelling used, the results depend on the parameter assignments chosen in the particular case. Conversely, the results can be transferred from simulation to the real world only if the particular parameter choices and assignments correspond to factual values, preferences, tendencies, or whatever. The value of simulation as a science of structure depends crucially on this factual question, even if many studies that work with such methods do not mention this point.

This raises a fundamental problem of validation. A paramount capacity for prediction or imitation, as in atmospheric circulation, cannot be discerned easily in the case of social simulations. Also, there are no arguments supporting why the specific parameter assignments correspond to the real social world. Furthermore, most of these assignments are made according to a backwards logic; that is, they are oriented toward performance in an instrumentalist way. What remains is a third strategy that argues with robustness, as Schelling did, to show that the exact parameter values do not matter. I do not intend to discuss in detail how one could validate such simulations and how one could gain knowledge from them. These simulations originated from the observation that a remarkably simple configuration can generate certain patterns. I shall be content with pointing out that the dilemma of plasticity is tantamount to a problem for validation.

Some scholars aim to solve the problem by propagating a new explanatory model. "Can you grow it?" is the crucial question in Epstein and Axtell (1996), meaning that a full-fledged explanation of a phenomenon would be the ability to generate it "from below." On the one hand, this approach introduces a promising perspective; on the other hand, it seems obvious that simulation fulfills this criterion more or less by definition, because once the process of specification is finished, the mechanisms implemented in the simulation model are able to generate the targeted phenomenon.[12]

12. Wise (2004) points to a broader movement in the sciences that aims toward such genetic explanations.

We can subsume the galaxy example discussed earlier: the result is yes, one can grow forms of galaxies from CA models. Nonetheless, I would insist on adding that the explanatory potential depends primarily on the relationship between the CA model and the physical theory about the formation of galaxies. It would be a fallacy to take the adequacy of patterns as evidence for the explanatory virtue of the generating model's structure, because the patterns have been produced by structure *and* specification.

CONCLUSION: PLASTICITY AND SIMULATION

A closely related problem regarding the relationship between computer and simulation was already debated intensely three decades ago. The starting point for this debate was the rather theoretical discourse around the computer as a universal Turing machine. All actions that can be described precisely can also be performed by a suitably programmed computer. This is the message of the Turing–Church thesis. In a way, it says that the computer as a machine has universal plasticity. Precisely because the computer can be adapted to extremely varied tasks, its structure is practically unrelated to those tasks. Robert Rosen (1988) elevated this observation to an essential characteristic of simulation in the following way:

> All relevant features of a material system, and of the model into which it was originally encoded, are to be expressed as input strings to be processed by a machine whose structure itself encodes nothing. That is to say, the rules governing the operation of these machines, and hence the entire inferential structure of the string-processing systems themselves, have no relation at all to the material system being encoded. The only requirement is that the requisite commutativity hold . . . between the encoding on input strings and the decoding of the resultant output strings. . . . This is the essence of simulation. (p. 532)

In a certain way, Rosen is right, but he chooses a rather abstract view of the machine. One important issue necessarily escapes such a perspective. In the present work, I want to view simulation from a different angle. From a logical point of view, computers can do many things in *some* way. I take this for granted, but it is of minor interest. The relevant issue from a philosophy of science point of view is (or so I see it): *How* does simulation modeling actually achieve what it achieves and how can it be characterized? This is a much more specific question.

At this point, it is appropriate to refer to the new robotics or behavior-based artificial intelligence. This is a multidisciplinary line of development that does not pick up the specific qualities of simulation as mathematical modeling but, rather, starts from the machine side. It aims at imitating certain types of behavior (catching something, walking, orienting in space) and hopes to overcome the (allegedly overly) general conceptual framework of traditional artificial intelligence. It was the strong connection to language, according to the proponents of the new approach, that hampered decisive progress in artificial intelligence (see, e.g., Brooks 1991, 2002; Pfeifer & Scheier, 1999).

Similar questions emerged in the context of computer science when the behavior of a software program had to be related to the hardware structure that it was built upon. Here, too, the concept of specification emerges that I was well aware of when choosing the same word for the process in simulation modeling. Let me quote Christopher Langton (1996):

> We need to separate the notion of a formal specification of a machine—that is, a specification of the logical structure of the machine—from the notion of a formal specification of a machine's behavior—that is, a specification of the sequence of transitions that the machine will undergo. In general, we cannot derive behaviors from structure, nor can we derive structure from behaviors. (p. 47)

The missing derivational relation is an important limitation, especially because this is addressing very formal processes that apparently lend themselves to a mathematized treatment. Michael Mahoney has contributed a series of historical and philosophical studies that explore the failure or limits of mathematization in the case of computer science (see, e.g., Mahoney, 1992, 2010).

Let me come back to simulation modeling. This chapter showed quite plainly how plasticity and the exploratory mode complement one another and are even dependent on each other. Without plasticity, simulation modeling would consist merely in a pragmatically strengthened or boosted variant of traditional mathematical modeling. With plasticity, it is even to be expected that models possessing high plasticity will call for artificial components—in particular, nonrepresentive artificial elements that are included to attain the desired performance. Plasticity requires specification, and only an exploratory mode makes models of high plasticity usable. This observation about structural underdetermination, however, does not imply that nonreproducibility of the target structure is among the general aims of simulation modeling. Instead, it tends to be associated with those practical strategies that enable simulation modeling to deal with situations that are highly nonideal.

4

Epistemic Opacity

In a passage from his memoirs, Richard Feynman (1985), the famous physicist and Nobel laureate, describes how he once succeeded in astounding a colleague. This colleague had approached him with a complicated mathematical calculation problem, but instead of ruminating on it and searching for a solution, Feynman simply told him the answer straight away. The colleague suspected that, somehow or other, Feynman must have known the result by chance. So he asked Feynman for further estimations that would normally require the use of a calculating machine. Again, Feynman took only seconds to give the correct answers. The colleague could not understand it. It would have been far too much of a coincidence for Feynman to have known all these results by chance; on the other hand, it was hard to believe that he possessed the seemingly superhuman capability to estimate complicated problems to an accuracy of several digits. Frustrated, he ended the conversation.

In his book, Feynman grants a look behind the curtain and explains the formidable mathematical capabilities for which he was so renowned. Apparently, he was too proud of them and of his fame as a wizard to offer his colleague an explanation. In fact, the prowess in this story is more about imitating mathematical computing power, because Feynman had found instant ways to treat the problems by performing approximations with the help of known results, and this was something that his colleague would also have been quite able to apply. Feynman consciously imitated the "number cruncher" and dealt in a playful way with the inferiority of human beings when it comes to calculation. One could say that Feynman was able to walk on water because he knew where the steppingstones were. Naturally, in the aforementioned episode, he had been lucky to be presented only with problems that did allow such shortcuts.

This is just one anecdote showing how Feynman was deeply convinced that humans cannot compete with computers when it comes to calculation, but that analytical understanding is still superior. The computer calculates much more quickly, but to look through something and to make a situation intelligible (when spiced with genius) is also an efficient way to lead to results—and because of the understanding involved, these results are potentially the more valuable ones.

However, this episode took place at a time when computers and especially simulation methods were much less advanced than they are today. Two things are different now: First, many practically relevant scientific problems arise in complex contexts. In such contexts, analytical understanding is often of only limited scope—the steppingstones are few and far between. Second, simulation modeling has developed into an instrument that is able to circumvent the complexity barrier—at least in a certain sense that I shall explicate later in this chapter. Simulation does not restore analytical understanding; on the contrary, the dynamics of the simulation models themselves remain opaque.

This does not necessarily lead to a loss of control over the model behavior, because simulations can compensate for the missing transparency with the help of the means discussed in the previous chapters: iteration, simulation experiment, and visualization. However, employing experimental approaches to become acquainted with the model behavior in a quasi-empirical way can substantially hinder the capacity for explanation. The previous chapter discussed how plasticity as a property of models can cast doubt on the explanatory capacity. The present chapter approaches the topic from the perspective of the modeler—that is, as an epistemic problem.

Epistemic transparency or lucidity can be an outstanding characteristic of mathematical models. Such models serve as surrogates for the sciences when they replace an impenetrable and complex world. In this respect, simulation models differ starkly from traditional mathematical models. Therefore, epistemic opacity is in need of a detailed examination.[1]

In the following, I shall discuss a couple of examples in which epistemic opacity occurs in the context of simulations. Some of these examples have appeared earlier. In general, most of the examples I use for illustration also reveal traits discussed in other chapters. This is exactly what to expect when these examples are not isolated special cases but, rather, cases showing simulation modeling as a new conception of mathematical modeling.

Overall, it will become clear how simulation models can provide a means to control dynamics, even though these models are epistemically opaque. Researchers and practitioners can employ a series of iterated (experimental) runs of the model and visualizations in order to test how varying inputs are related to outputs. In this way, they can learn to orient themselves within the model—even if the dynamics of the simulation remain (at least partly) opaque.

Admittedly, such an acquaintance with the model falls short of the high epistemic standards usually ascribed to mathematical models. This lower standard is still sufficient, however, when the aim is controlled intervention in technological contexts. On the other hand, opacity has to be accepted if the option for control is to remain in any way open. This chapter accordingly closes by discussing whether epistemic opacity restricts simulation-based science to a pragmatic—"weak"—version of scientific understanding.

THE GREAT DELUGE

Since the invention of the electronic computer in the middle of the twentieth century, the speed of computation has increased enormously. Work that once had called for an expensive supercomputing machine is, only

1. At this point, I want to make sure that I am not being misunderstood: there is no guarantee that analytical solutions will be automatically tractable or even transparent—they do not need to be elementary (see Fillion & Bangu, 2015).

one decade later, nothing out of the ordinary for machines that can be purchased in just about any city in the world. However, mere calculating speed does not necessarily correspond to a capacity to solve problems. Alan Turing, who worked out a machine on paper that corresponds essentially to the modern computer, already gave us fundamental insights into computability and its limitations (see also Lenhard & Otte, 2005). The majority of real numbers, for example, cannot be computed—a fact that originates from the relationship between the abstract world of numbers and factual machines constructed according to a certain architecture. The question regarding which types of problem cannot be solved efficiently with the calculating speed of computers (of the Turing machine type) led to the explication of Kolmogorov–Chaitin complexity (see Chaitin, 1998). One also uses the term "computational complexity" to describe the sort of complexity that, so to speak, measures how well a computer can cope with processing a problem.

What makes this complexity so interesting is that a great number of problems emerge in scientific and technical practice that are complex in exactly this sense. I should like to take the well-known "traveling salesman problem" (TSP) as a prototype: A number of cities are given with the appropriate distances between them, and the problem is to find a route that links all the cities while simultaneously being as short as possible. For each such problem with n cities, there exists a solution—so much is obvious. Moreover, several routes may well provide the shortest traveling distance. However, how can one find such an optimal route? This question formulates a typical optimization problem that is in no way simple to solve—even with the enormous computing power available today. Why is this so? Eventually, it would not be all that difficult to work out an algorithm that would try out all the different routes, and (at least) one would have to be the shortest. This approach is correct *in principle,* but not feasible *in practice.*

The problem is that even a moderately large n will result in so many potential routes that working through them all would overwhelm even the most powerful computer. The computational effort as a function of n expands more rapidly than any polynomial in n, revealing that "brute force" cannot be the key to any practical solution of the problem (technically speaking, the TSP belongs to the class of NP-hard problems).

Such and similar optimization problems are widespread in technical practice. A number of scientific disciplines are involved in determining solutions that are capable of practical implementation—the computer as the instrument has to be fed, so to speak, from all sides. This includes not only numerical and combinatorial optimization and scientific computing but also informational and technological approaches in programming and implementation. On the mathematical level, there are a host of different approaches that, in part, include very advanced ideas (see Gutin & Punnen, 2002). In the following, I shall present and discuss the Great Deluge algorithm developed by Gunter Dueck (IBM Scientific Research Center Heidelberg; Dueck, 1993). This example seems so suitable for illustrating characteristics of simulation modeling because it does away almost completely with theoretical considerations, using a direct simulation-based approach instead.

The problem is to solve the following task: One has a mathematically defined space, such as the space containing the possible routes of the traveling salesman in the TSP. A point x in this space is assigned a specific value $F(x)$—in the TSP, the length of route x. A neighboring point x', which emerges through switching the sequence of two cities, has another length $F(x')$, and so forth. If one conceives the generally high-dimensional spaces as being projected onto a plane and assigns every point x the functional value $F(x)$, one obtains a sort of mountain range, and the task is then to find the highest peak (or the deepest valley, which in mathematical terms means only switching the sign). How can one find this peak, even though one knows nothing about the shape of the mountain range? One can assign the height to every specific point, although one does not have an overview of all points. Bearing in mind the brief discussion on computational complexity, one can say that in many practical tasks, there are simply far too many points for even a supercomputer to check through.

Were that otherwise, such optimization problems would indeed be solvable by straightforward computing power. Complexity prevents any brute-force strategy such as "faster and more of the same" from yielding results. This fact has to be factored in if one wants to characterize simulation modeling as a new type of mathematical modeling.

Back to the optimization problem: One suggestion would be to start somewhere with a random choice, and then simply go "uphill"[2]—that is, to check only the neighboring points $n(x)$ of x, and then, in the next step, to maximize $F(n[x])$. This algorithm would certainly lead to a peak, though without in any way guaranteeing that it would be the highest or even one of the high ones. This would be the case only if, by chance, the starting point were already on the slope of this high peak. Hence, although this is a quick procedure, it generally fails to deliver a good result.

The Great Deluge algorithm now proposes a surprisingly simple way of modifying this approach by taking the following simulated hike through the mountain range of solutions. The hiker starts somewhere at random as before, but does not go uphill continuously; every new step sets off in a randomly chosen direction. The hiker moves according to a random process that takes no account of the form of the mountains. Parallel to this, there is a second process that gave the algorithm its name—namely, a great deluge that is causing the water level to rise continuously. Only one restriction is imposed on the random hiker: to avoid getting wet feet. Hence, whenever the random generator proposes a step that would lead to a point below the water level, another, new step has to be simulated. Clearly, this simulated hike ends at a local maximum as soon as all neighboring points lie under water. Because the hike can also lead downhill, the starting point does not ascertain the final outcome in advance.

An obvious objection to such an algorithm is that the simulated hiker is bound to get stuck on one of the islands that will inevitably form at some point as a result of the "deluge." In other words, is there any argument to explain why the local maximum attained should be equal to or almost equal to the global maximum? A convincing mathematical argument is missing at this point—and this has led some theoretically oriented mathematicians to criticize the whole approach. However, practice (e.g., finding good placements for components of computer chips) shows that

2. This should read "downhill" because the proposed solutions become shorter. In a technical sense, changing the sign allows one to replace minimization with maximization. I am doing this in order to make the process easier to imagine.

the algorithm delivers surprisingly good results—something that its inventor also notes:

> Frankly speaking, we did not seriously expect this algorithm to function properly. What if we were to start in England or Helgoland? When the Great Deluge arrives, even the continents will turn into islands. The Great Deluge algorithm cannot escape from land that is surrounded by water and may therefore be cut off from every good solution.
>
> But, strangely enough, we have applied the Great Deluge algorithm to optimize production, for large-scale chip placements, and even for the most difficult industrial mixing problems, and we have calculated thousands of TSPs. Again, only a tiny program was required and the outcome was excellent in each instance. (Dueck, Scheuer, & Wallmeier, 1999, p. 28; my translation)

How can one be confident that the outcome actually will be "excellent"? A mathematical proof of how good the algorithm works is not available. Being able to improve a solution that already works in industry would indicate that the algorithm has done a good job. A most comfortable judgment situation is when different approaches can be compared. Consider the example (an often-analyzed test case) of locating holes with a drill in a circuit board. The drill bit has to be placed successively at all locations of holes; hence, it proceeds like the traveling salesman who starts and finishes in a fixed town (the null position of the drill bit). Finding a good solution for this TSP is of obvious economic benefit, because it will raise the number of items produced per time unit. Figure 4.1 shows the circuit board and locations of all the holes to drill (towns to visit). Figure 4.2 shows the starting point of the simulated route—that is, a random path connecting all the holes (towns).

Figure 4.3 shows an intermediate stage; figure 4.4, shows the end result of the Great Deluge algorithm. It took about 450,000 steps, or a few seconds of computing time, and produced a path with a length of about 50.98 inches. In this special case, the optimal result is known from theoretical analyses performed by Holland (1987). It is 50.69 inches.

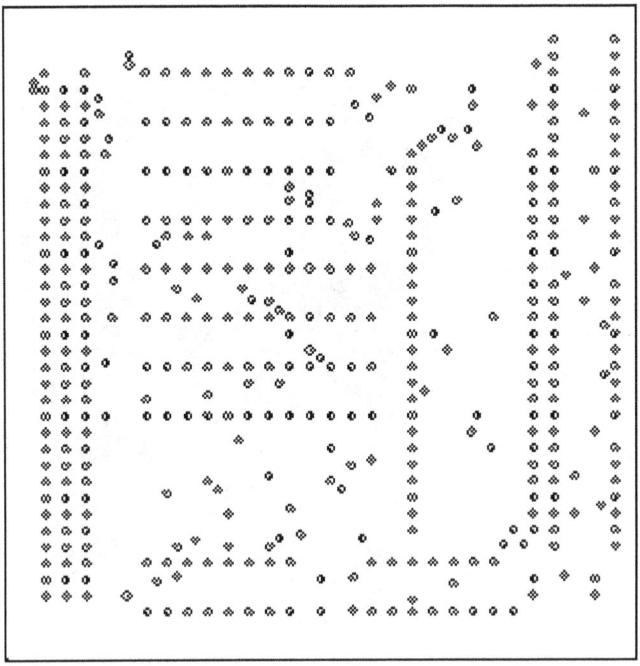

Figure 4.1 Circuit board with location of holes to be drilled. *Credit: Courtesy of G. Dueck.*

Though this success seems astonishing, it is only one example documenting the power of the algorithm—but it neither explains it nor guarantees its success. Obviously, problems in high-dimensional spaces do not just bring disadvantages, owing to reasons of complexity; they can also bring advantages. For instance, algorithms can succeed in high dimensions that would fail in three dimensions (already because of islands, as pointed out by Dueck, Scheuer, & Wallmeier, 1999).

Even the concept of an island entails that this object is isolated. One can interpret the success of the Great Deluge algorithm in the following way: islands are rare in high-dimensional spaces because such spaces possess a higher degree of connectivity than our three-dimensional intuition is willing to assume.

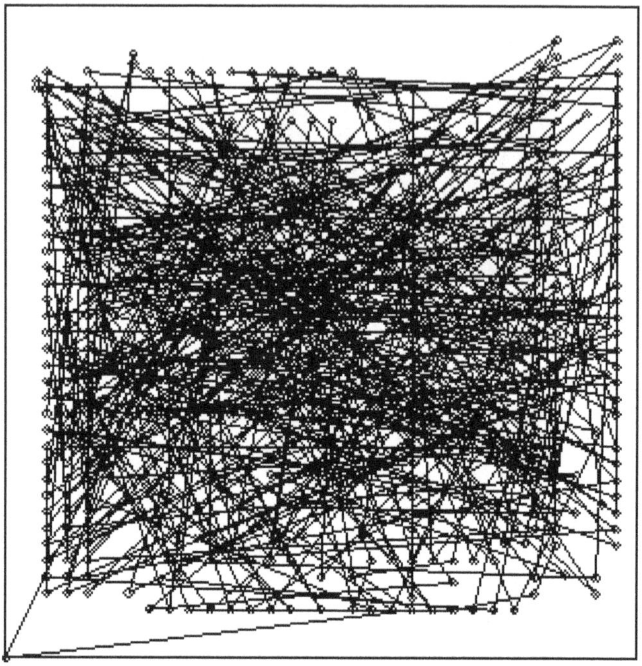

Figure 4.2 Initial configuration of Great Deluge algorithm. *Credit: Courtesy of G. Dueck.*

THE SURPRISING FORMATION OF A GOLDEN WIRE

I shall enlarge the pool of illustrative examples before moving on to a general conclusion. Two works by Uzi Landman, director of the Center for Computational Materials Science at Georgia Institute of Technology, Atlanta, will serve as my second and third cases. Landman is a pioneer of simulation methods in the materials sciences or, more precisely, in that part of them that belongs to nanoscale research.[3]

3. Regarding the disciplinary idiosyncrasies of materials research, see Bensaude-Vincent (2001). On the problems of taking simulation as the characteristic trait of *computational nanotechnology*, see Johnson (2006b).

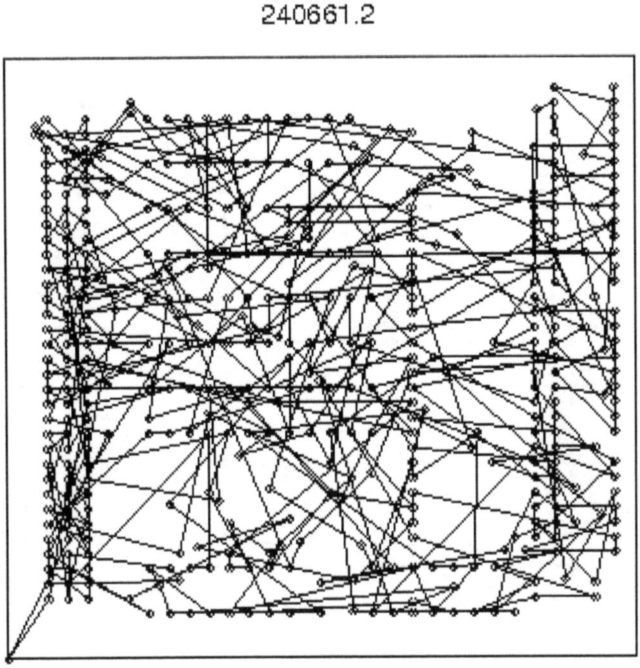

Figure 4.3 Intermediate stage of the Great Deluge algorithm. *Credit: Courtesy of G. Dueck.*

While working with molecular dynamics simulation (Landman et al., 1990), Landman was surprised by a spectacular observation. In this simulation, he and his coworkers brought a nickel tip as used in what are termed "atomic-force microscopes" into close proximity to a gold surface. They even let the tip make contact with the surface—something that experimenters normally want to avoid or regard as an accident. Subsequently, they slowly withdrew the (simulated) tip from the surface material. After creating this unusual situation, Landman was amazed to observe how gold atoms detached themselves from the surface and built a very thin golden nanowire connecting the surface with the tip.

Figure 4.5 displays six simulated snapshots of this process. In the upper left corner, the tip has been pushed into the surface. On the following images, it is being slowly withdrawn, thereby creating a nanowire—that is,

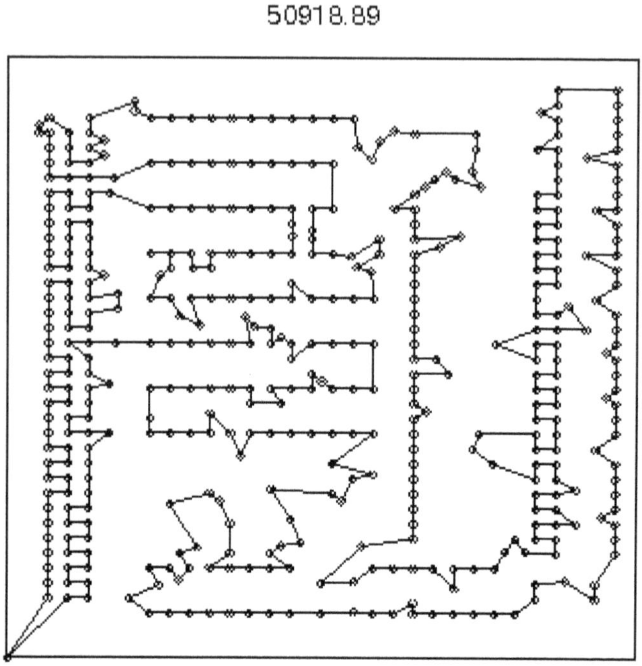

Figure 4.4 Final stage of the Great Deluge algorithm. *Credit: Courtesy of G. Dueck.*

a wire made of gold of only few atoms diameter. The images show a highly idealized way of using forms and colors; they can even be viewed as an animated sequence.[4]

The artificial coloring that makes it possible to grasp the dynamics of the several molecular layers contributes to making the sequence stand out. Landman, in an interview, compared his own situation to that of a researcher observing the outcome of a complicated experimental setting:

> To our amazement, we found the gold atoms jumping to contact the nickel probe at short distances. Then we did simulations in which we withdrew the tip after contact and found that a nanometer-sized

4. Rohrlich (1991) gave this sequence as an example for simulations being "dynamically *anschaulich*."

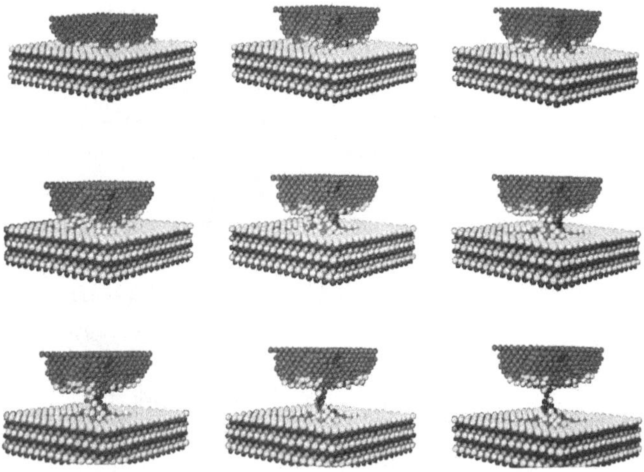

Figure 4.5 A golden nanowire emerges between a nickel tip and a gold surface. *Credit: Courtesy of U. Landman.*

> wire made of gold was created. That gold would deform in this manner amazed us, because gold is not supposed to do this. (Landman, 2001)

The outcome was surprising, because—contrary to expectations—gold was chemically active instead of remaining inert. Of course, one should bear in mind that everything was happening in a simulation and was observed only as software output on a screen. Behind the simulated phenomenon there were no empirically measured data in the normal sense. Instead, there was a massive employment of computers to simulate the interactions between the atoms via molecular-dynamics methods. By the way, the figure shown here has been polished to maximize its intuitive impact. Georgia Tech has its own special department for supporting visualizations. Hence, this case could have been used just as much to explicate exploratory experiments and visualizations if our focus had been on how Landman calibrated the model and how the published figure was produced. Plasticity and artificial components also play an important role in methods of molecular-dynamic modeling. However, analyzing these methods would call for a detailed discussion of the algorithms used in the modeling process. At this point, I shall dispense with this analysis.[5]

In sum, a new and surprising phenomenon was created in simulation—namely, the emergence of a golden nano-sized wire. The simulated process appears highly graspable, but this does not guarantee the phenomenon has more than "virtual" reality. Does the phenomenon also occur in the natural world, or is it an artifact of a misleading model world?

In fact, the simulated effect turned out to be a correct prediction when two years later, empirical results of atomic-force microscopy confirmed the phenomenon.[6] The amazement that Landman reported turns up again on a theoretical level: whereas well-known physical laws and well-calibrated forces served as the starting point for the simulation, the further dynamics led to unexpected behavior. The local theoretical descriptions of interactions did not render transparent the behavior in the larger molecular compound.

LUBRICANTS BETWEEN SOLID AND FLUID

The third example also comes from Landman's simulation laboratory and, as can be expected, is very similar in appearance. It treats a problem of tribology on the nanoscale—that is, the properties of lubricants that have to function in very narrowly confined spaces such as a gap of just a few atoms, a "nano-gap" (Gao, Luedtke, & Landman, 1998). Richard Feynman has been heralded as a visionary of nanoscale research since

5. Frenkel points out that the calibration of force fields presents a central and at the same time tricky challenge for molecular-dynamics methods (Frenkel, 2013, sec. 2.3). I would like to refer also to Scerri (2004) and Lenhard (2014), who discuss artificial elements, tailor-made for simulation purposes, in *ab initio* methods of quantum chemistry. Johnson (2009) differentiates *computational nanotechnology* according to the simulation methods used. In particular, she distinguishes theory-oriented from technology-oriented approaches, in which the latter utilize plasticity without reservations.

6. The methods of atomic-force microscopy also rely on using computational models. Even if what is visible on a computer screen does not relate directly to what the tip of the microscope measures, it is still an essentially empirical input.

his programmatic lecture "There Is Plenty of Room at the Bottom (1960). Already in this lecture he had supposed that friction and lubrication would present new and unexpected phenomena on the nanoscale. Feynman relied on abstract theoretical considerations— namely, that the small nanoparticles have, in relation to their volume, a much larger surface than chunks of material in classical physics. It is typical that Feynman could come up with correct theoretical reasoning and anticipate new phenomena in friction and lubrication while, at the same time, being unable to describe or even predict specific phenomena. Feynman himself was entirely clear about that: his lecture was on room for new phenomena, not on new phenomena themselves. Statements about the behavior of lubricant molecules that are, at least potentially, relevant for technological applications are hardly possible, for reasons of complexity. These reasons will be explained in more detail later. At this point, complexity erects a real barrier against understanding.

Landman's simulations, however, provide concrete instances for Feynman's conjectures. These simulations suggest that the behavior of long-chain lubricant molecules resembles that of both solids and fluids at the same time—what Landman calls "soft-solids."

Figure 4.6. displays the outcome of a simulation experiment in which two layers slide one against the other. The gap between the layers contains a thin film of lubricants that is only a few molecules thick. This is fed by a larger amount of lubricant outside the layers.

The upper part of figure 4.6 displays a (simulated) snapshot of the sliding layers. In the gap, the lubricant molecules build ordered layers, and this strongly increases the friction of sliding. Now, in the simulation, all molecules in the gap become highlighted (in black and white print: dark). Landman and his colleagues then tried to overcome the problem of high friction created by ordered layers. They added a small oscillation to the movement, let the movement go on for a while, and then took the second snapshot, depicted in the lower part of figure 4.6. Indeed, this visualization indicates the success of the oscillation, insofar as the lubricant molecules have lost their order. Some of the marked molecules that had been stuck in the gap have now moved outside and have mixed with other lubricant molecules. These *soft-solid* properties, as Landman called them, are surprising compared to the normal behavior of either

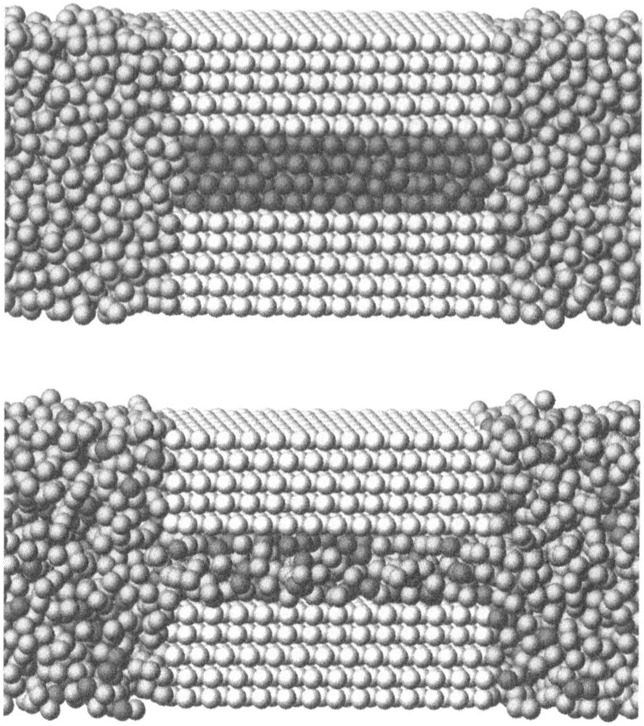

Figure 4.6 Ordered state and high friction (above); low order and low friction induced by oscillation (below). *Credit: Courtesy of U. Landman.*

fluids or solids: "We are accumulating more and more evidence that such confined fluids behave in ways that are very different from bulk ones, and there is no way to extrapolate the behavior from the large scale to the very small (Landman, 2005).

In 2002, Landman was awarded a medal by the Materials Research Society (MRS) for his simulation studies on how lubricants behave. Landman is a materials researcher rather than theoretical physicist—which makes a significant difference. The former is oriented toward technologies, and therefore control of phenomena and their quantitative characteristics are key. If options for interventions become available, theoretical understanding becomes dispensable.

SURPRISE, CONTROL, AND THE COMPLEXITY BARRIER

Perhaps the first thing to attract attention is that all three examples described here contain an element of surprise. This is a quality that has always been important in philosophy: science should not operate dogmatically, but should be open to surprises. Normally, however, this demand is associated with an empirical component: nature is not transparent and sometimes does not follow our expectations. At first sight therefore it might appear surprising that *simulations* should also exhibit this property. Although simulation models have been constructed, their behavior does not need to be in any way transparent. This may be due to the complexity of the inner dynamics or to the complexity of the conditions that the construction has to respect when, for example, different groups of researchers work on different modules, or when significant parts of the model have been purchased as commercial software. I shall address this aspect in chapter 7, when considering the limits of analysis. The present chapter deals mainly with the first aspect: simulation dynamics can hold surprises—can be epistemically opaque—even if each single step seems to be completely transparent.

The Great Deluge algorithm gets by with almost no theoretical assumptions. Most methods of optimization are based on theoretically more elaborate approaches that, as a rule, utilize the special structure inherent in the particular problem. For instance, there are advanced methods for making an optimal choice of the grid, meaning that the simulation should attain a balance between precision and efficiency. The Great Deluge algorithm does not take this into consideration. On the one hand, the simulation, in lieu of theoretical structure, appears to be similar to a cheap trick. On the other hand, this is what grants it a remarkable scope of application. The surprising element consisted in the observation, as the astonished inventor expressed in the earlier quote, that the method works reasonably well at all. Evidently, the objection that the simulated hiker would get stuck on an island has no power. The simulated movements teach us not to trust our intuitions in higher dimensions.

The simulated hiker who repeatedly arrives at peaks of similar height acts like a kind of proxy for humans and so gathers experiences in high-dimensional topography. This reveals that the condition to stay "dry" is not a strong condition; in high-dimensional spaces, there are many pathways.

What both simulation examples from the nanoworld have in common is that unknown and surprising phenomena occur, although the fundamental laws governing these areas (mainly quantum theory; molecular models work with force fields) are known. This fact probably contributes to the fascination of nanoscale research when it ventures to discover previously unseen phenomena and novel properties of materials. There is no contradiction involved, because the theory does not entail statements about how materials behave under particular conditions.

An early, famous, and similar example is the planetary motions in our solar system. The planets influence each other via gravitation, and hence a small change in the movement of one planetary body could possibly lead to a significant change in the long-term behavior of the entire system. Henri Poincaré found out, in an ingenious but laborious way, that even the dynamics of three (planetary) bodies—what is termed the three-body problem—cannot be integrated mathematically (analytically). The question whether our solar system is stable is closely related and cannot be answered for reasons of complexity—at least not with the analytical apparatus of mathematics. Thus one can call this a "complexity barrier." Cases such as nanoscale research are similar insofar as theories of local interactions are accepted, whereas the derivation of the interesting global behavior is blocked.

Hence, the problem of the complexity barrier is present in all the examples from nanoscale research. The Landman quote given earlier that "there is no way to extrapolate behavior" was pointing to the difference between classical physics and nanophysics. The laws of large numbers at work in the classical (continuum) case do not hold. The difficulty in extrapolating model behavior is effective in the opposite direction, as well. The dynamics between atoms are formulated theoretically—and nobody doubts the validity of the Schrödinger equation. However, it cannot be used to derive relevant facts about phenomena of friction of the kind simulated by Landman. Only the (simulated) interaction of components brings out the model dynamics (the case of atmospheric circulation discussed in the first chapter is also similar to this).

In all three examples discussed here, simulation methods led to remarkable results. This poses a twofold question: How does simulation overcome the complexity barrier? And what does the answer to this question imply when it comes to the conception of mathematical modeling?

My argumentation comes down to two claims: First, simulations can facilitate orientation in model behavior even when the model dynamics themselves remain (partially) opaque. And second, simulations change mathematical modeling in an important way: theory-based understanding and epistemic transparency take a back seat, whereas a type of simulation-based understanding comes to the fore that is oriented toward intervention and prediction rather than theoretical explanation. Thus, simulations circumvent the complexity barrier rather than remove it. They present a kind of surrogate for frustrated analytical understanding. This finding gives simulation modeling an "instrumental" touch; but simulation does not command a form of instrumentalism. Or, more precisely, simulation does not fit well in the old divide between instrumentalist and realist positions. For one reason, theories play too strong a role. Simulation rather undermines the divide. I shall discuss this point in my final chapter 8.

SIMULATION AS A PROGRAM FOR SUBSTITUTION

Humphreys rightly emphasizes opacity as a characteristic trait, or typical danger, of simulations (briefly in 2004 and decidedly in 2009). However, complexity barriers appear again and again in the history of theory building and mathematical modeling, and the opacity issue is not exclusive to computer modeling. In a first step, I shall briefly discuss cases in which such barriers have occurred, and then consider which strategies have been utilized to overcome them. In a second step, I shall then compare these cases to simulation modeling.

One famous example is the introduction of algebra in the sixteenth century. In Europe, this new instrument triggered major advances in the practice and theory of arithmetic. These developments served as the role model for the differential calculus that the young Leibniz was aiming to create. He set himself an explicit goal: to create an instrument for treating geometrical-analytical infinitesimal problems that would be as efficient as calculating with letters was—the instrument Viète had formed for algebraic problems (Hofmann, 1974). In both cases—algebra and calculus—the instrumental aspect was in the forefront and decisively influenced the dynamics of mathematization.

Based on the Newtonian theory of gravitation, infinitesimal and integral calculus had become the backbone of rational mechanics during the eighteenth century, and its agenda dominated the (mathematized) sciences. This seemed to have brought mathematics close to completion. An eloquent exponent of this opinion was Diderot, whose Thoughts on the Interpretation of Nature from 1754 (Diderot, [1754]1999) foretold that the next century would hardly see three mathematicians of rank equal to such contemporary heroes as Leonhard Euler, the Bernoullis, Jean-Baptiste le Rond d'Alembert, and Joseph-Louis Lagrange. A prediction, by the way, with which Lagrange agreed.

Pierre-Simon Laplace's famous image of a "demon" also points in this direction. The universe is determined completely by its mathematical description; however, one would need the computational capabilities of a godlike demon to assign the exactly correct initial conditions and to actually calculate what follows from this mathematical description. Hence, the universe was conceived of as theoretically and mathematically fixed, whereas it could not be, or could merely rudimentarily be, deciphered on a practical level. In complex systems, it is hard to extract predictions from the mathematical description that are more than insufficient guesses. As a predictive instrument, mathematics fails.

Conceptual change had no place in this picture, but was all the more striking when it happened—as the relationship between Euler and Évariste Galois shows.[7] Leonhard Euler (1707–1783) was one of the leading scientists and arguably the most prominent mathematician of his time—a widely acknowledged master of analysis (differential and integral calculus) that he forged into the central instrument of rational mechanics. Many of

7. The choice of these persons is not without alternatives. Considering Carl Friedrich Gauss as a progressive inventor, for instance, would lead to similar observations.

Euler's results sprang from tedious calculations spread over several pages that showed such a degree of complication and technical mastership that many contemporaries—including Diderot, as we have seen—gained the impression that nobody could possibly make further progress in this field. At the turn of the eighteenth century, the whole field seemed to have reached a stalemate, because Euler's works had already pushed forbidding opacity to its limits. The young genius Évariste Galois (1811–1832) criticized Euler's methods precisely because they left no room for progress. Galois (1962) viewed conceptual change as the only alternative.

Algebraic analysis presented a conceptual innovation—namely, a kind of arithmetization that made algebra directly applicable to mathematical relations (see Jahnke & Otte, 1981). This move toward abstraction turned mathematical facts and relations into objects for mathematization, and thus enabled the dynamic development of mathematics that characterized the nineteenth century in such a remarkable way. This strategy used abstraction to introduce a new level of objects, with the goal of achieving tractability for a field lost in complexities. It was a fundamental source of mathematical development (Otte, 1994). At the same time, Galois's turn renewed the instrumental virtue of mathematization, because it created a new pathway to remove opacity.

Michael Detlefsen (1986) located Hilbert's program precisely in this line: it should solve the foundational problems of mathematics with the help of the proof theory that Hilbert had devised for this purpose. The basic idea was to formalize ways of mathematical reasoning so that one would then be able to investigate them on a more abstract level. Intuitionism, for instance, required (in disagreement with Hilbert) working without what has been termed actual infinite, because it could not be used in a reliable way by human beings for epistemological reasons. Hilbert dissented here, favoring a solution that has been called formalistic because it conceives the relevant activities of reasoning as manipulations of formulas. According to Hilbert's view, it is sufficient for such a calculus to be consistent (free of contradictions). On the level of (logical) calculus, the concern is not a substantial justification of reasoning with infinities but, rather, a formal proof of logical consistency. The manipulation of objects that represent infinities, so Hilbert's point, can itself work in a completely finite manner.

For Detlefsen, it is adjoining "ideal elements," which are oriented toward the properties of the formal system and optimize its instrumental virtues, that deliver the core of Hilbert's "instrumentalism." As in the examples from the history of mathematics discussed earlier, Hilbert's method of ideal elements is—according to Detlefsen—an instrumental shortcut in methods of reasoning. This shortcut generates epistemic transparency, though on a new, more general, and more abstract level. The new method was developed to counter the threat of ending up in a stalemate of inefficient operations. Detlefsen describes this with the apt expression of circumventing "epistemically paralyzing length or complexity" (Galois expressed himself differently, but with a similar meaning):

> On this account, it is its ability to simplify which is the source of the epistemic efficiency of the ideal method. . . . In a nutshell, the use of ideal elements widens the scope of our methods of epistemic acquisition without complicating those methods to such an extent as would make them humanly unfeasible. . . . More generally, our ability to acquire epistemic goods is enhanced by our having an "algebraic" or "calculatory" representation of the world, which allows us to circumvent contentual reasoning which can be of an epistemically paralyzing length or complexity. (Detlefsen, 1986, p. 9)

Hence, there is an established discourse in the history and philosophy of mathematics that has epistemic opacity as its subject—at least somewhat implicitly—when discussing ways and means to increase transparency. Nonetheless, it should not be forgotten that Hilbert's program is based to a high degree on a new *conceptual* formation, insofar as the virtues of an axiomatic system become a perspective in their own right (see Lenhard & Otte, 2005). Thus a twofold perspective is needed: both conceptual and instrumental aspects have to interact in order to bring about the dynamics in the course of which relations and ways of reasoning become mathematized and then develop their own life.

What happens if the situation becomes opaque? In an epistemic respect, the goal is (also) to reconstitute analytical understanding. Richard Feynman has expressed this kind of understanding in a concise way (the first sentence cites Paul Dirac):

> I understand what an equation means if I have a way of figuring out the characteristics of its solution without actually solving it. So if we have a way of knowing what should happen in given circumstances without actually solving the equations, then we "understand" the equations, as applied to these circumstances. (Feynman, Leighton, & Sands, 1965, vol. 2, sec. 2-1)

In fact, this understanding proves its analytic quality by rendering lengthy and complicated reasoning procedures obsolete. In short, it presents the counterpart to epistemic opacity. With a pinch of salt, one could state: mathematical modeling is performed in the service of epistemic transparency. This does not mean that mathematical models are epistemically transparent per se. On the contrary, their development has been plagued with opacity. Even analytical solutions can be intractable.

Against this background, simulation modeling adds a radically new twist to mathematics: simulation does not care about epistemic transparency and therefore cannot generate the aforementioned analytical understanding. Instead, simulation exploits its instrumental virtues by managing with less—that is, with a surrogate for analytical understanding.

Quite in contrast to what Feynman had in mind, simulation can provide an orientation toward the behavior of models that comes from experience with iterated calculations as found in the variation of parameters. Viewed in this way, simulation can indeed offer a way of knowing what should happen in given circumstances, although this way of knowing is not based on theoretical insight. One could formulate the recipe in the following way: Simulate and observe the results, iterate the procedure in order to explore the model dynamics, and adapt the model in an appropriate way in order to finally obtain an overview of model behavior. Nota bene, the latter presents only a surrogate for theoretical understanding. In other words, it attains a comparable effect on an essentially different basis.

Researchers can orient themselves toward the behavior of a model by first exploring and then acquainting themselves with its behavioral patterns—even if their orientation is not based on analyzing, or making transparent, the model structure. This can count as a surrogate, because becoming acquainted with model behavior delivers options for controlling the generated phenomena, as shown by the given examples: Which kind of oscillation, for instance, is appropriate to reduce friction? Options open up for control, intervention, and prediction. This makes simulations into an instrument tailor-made for the requirements of a science oriented toward technology.

Nevertheless, a surrogate is merely a surrogate. The superior computational capacity of the computer is indispensable for simulation when a relationship has to be investigated that connects varying model assumptions with the model behavior resulting from them.

Delegating these computations to the machine is instrumentally effective, but it also renders all the following single steps incomprehensible. Thus, epistemic opacity is part and parcel of simulation modeling.

EPISTEMIC OPACITY

Indeed, epistemic opacity crops up constantly as an issue when discussing examples of simulation modeling. It emerges for the following reasons:

1. The sheer number of computational steps renders the entirety of a computation unsurveyable. It is primarily this aspect that Humphreys (2004, p. 148) emphasizes when speaking of epistemic opacity. He is correct, but there are further reasons as well.
2. The discrete nature of computer models requires artificial components to balance the overall behavior of the models (as discussed in chapter 1). These components serve as measures for controlling and adjusting the model implemented in the computer, even if this model deviates from the dynamics defined in theoretical terms. However, only the latter would be comprehensible, and this is why artificial components and epistemic opacity correlate positively.
3. The plasticity of models was addressed in chapter 3. Using tuning knobs to calibrate the simulated dynamics according to global performance measures fosters opacity. In general, the "plastic" portion of a model, which is specified in an iterative–exploratory manner, will lead to opacity.

4. I would like to recall the case of Hurricane Opal, in which different parameterization schemas were tested with the help of visualizations. Although the internal model dynamics to a large part remained opaque, the behavior of the modeled hurricane could be controlled; that is, it could be adapted successively to the preassigned trajectory. It was not just the mere number of relevant variables that made the model dynamics opaque. The fact that independently developed simulation modules were coupled together was of equal significance. Opal, or its model, was based on an atmospheric model taken from the shelf of another institution. The modularity of complex models, especially ready-made plug-in modules, contributes to opacity. The way simulation modeling is organized thus leads to opacity.

Hence, there is a series of reasons why epistemic opacity appears in the context of simulation. The concept of opacity and its systematic relevance have yet to be appreciated adequately in the philosophy of science literature. Nonetheless, there are exceptions, and Paul Humphreys deserves first mention. He considers opacity as being a systematic (if unwanted) companion of simulation (2004, p. 148; and 2009, sec. 3.1). I fully agree with him. But, in addition, I would like to highlight the role that opacity plays in characterizing simulation.

Opacity is not just an epistemologically deplorable side effect; it is also the direct downside of simulation's instrumental strengths. The inclusion of artificial components and the specification phase of plastic models compel opacity.

Although one should not underestimate the role of opacity in simulation, it is not the only significant and determining factor. In this respect, simulation is just one element in a series of mathematical techniques, as has been discussed. As a machine, the computer requires detailed instructions. What appears, on the level of visualizations, to be a more or less informative output, has been generated by software that respects very high consistency standards—otherwise the software programs could not be compiled. These conditions can be met only with the help of semantically advanced programming languages such as Fortran or C. Such languages can themselves count as instruments for overcoming those opacity-related obstacles that would be present on the semantically lower level of machine language.[8]

An important part of computer science is devoted to developing adequate instruments for a mathematization that will render complex and unclear computer processes theoretically treatable. Michael Mahoney (1992, 1997, 2010) has contributed a series of historical and philosophical analyses of computer science and informatics in which he investigates the attempts to mathematize informatics. Such attempts, as is the tenor of his studies, had a very plausible starting point, because the computer is (or appears to be) a formally determined machine. Mahoney's diagnoses, however, amount to seeing theoretical informatics fail in its attempts to mathematize. What Mahoney describes as the failure of (traditional) mathematical modeling is perhaps better captured as a new type of mathematical modeling that allows us to deal with epistemic opacity—if the latter cannot be made to disappear.

Di Paolo, Noble, and Bullock (2000) refer to opacity when they describe simulations as opaque thought experiments. Though I find the context of thought experiments too narrow when it comes to investigating simulations, the authors observe correctly that there is "a role for simulation models as opaque thought experiments, that is, thought experiments in which the consequences follow from the premises, but in a non-obvious manner which must be revealed through systematic inquiry" (p. 1). In mathematical models, it is quite typical for the accepted assumptions entered into the model to interact in unforeseen ways and lead to unexpected consequences. In general, I would like to argue, mathematical techniques serve to derive consequences where mere thought experiments do not succeed.[9] Simulation offers an opportunity for "systematic inquiry" that remains valid even under the condition of partial opacity.

In the philosophy of mathematics, the employment of computers was first discussed in the context of computer-assisted proofs. The computer-assisted proof of the four-color theorem (Appel, Haken, & Koch, 1977;

8. On the development of the languages C and C++ and their role as lingua franca, see Shinn (2006).

9. See Lenhard (2017a), who discusses what that may mean for the relation between simulation and thought experiments.

see also Appel and Haken 1977) has been debated critically, particularly because of the way it distinguishes among so many cases that only a computer can manage it. At the time, it was considered scandalous that such an epistemically paralyzing length or complexity (to use Detlefsen's apt expression) should appear in a mathematical proof, because proofs were deemed to present the paradigms of epistemic lucidity. Tymoczko (1979) has pointed to *unsurveyability* as the philosophically central trait of this new kind of computer proof.

I have shown here, when discussing the cases of Galois and Hilbert, that *unsurveyability* is not related as exclusively to the computer as Tymoczko assumes. A similar diagnosis applies to the proof that Andrew Wiles delivered for Fermat's theorem. A single person, given superior talent and education, would need years to retrace this proof thoroughly. The particularity of proofs such as that of Wiles is that they proceed to the limit of what is intersubjectively testable in a scientific discipline.

Some of the relevant issues when using computers in mathematics are well known from the more general context of technological instruments in the sciences. A bit earlier, I discussed the paradox Blumenberg (1965) mentioned in his work on Galileo: although instruments amend and reinforce the human senses, the results are then marked by the contingencies of the instruments as well. With computers and simulations, this observation now also applies to mathematical modeling. However, there is still conceptual novelty in this development, because mathematics and mathematical modeling are commonly conceived as playing the counterpart to instrumental and empirical components in the sciences. They are now losing this role. Valentin Braitenberg, for instance, diagnoses that mathematics is making itself independent. What he means by this is that simulation models do not merely serve to illuminate an intransparent nature, but may *themselves* be complex and opaque objects. "It is shattering that mathematical theory, when making itself independent in the computer, dethrones the unquestioned and highest authority for finding the truth, namely logically educated human reasoning" (Braitenberg & Hosp, 1995, p. 9; my translation).

Some diagnoses, though correct, are motivated presumably by the hope of turning back the clock. Frigg and Hartmann (2008), for — instance, warn against a seductive property of simulation—namely, that simulations encourage continuously enlarged and complicated models that will eventually become incomprehensible. To a certain extent they are right, but they suggest that opacity might be prevented by conceptional rigidity. At this point, I disagree. I consider opacity to be part and parcel of simulation modeling—or, expressed less strongly, preventing opacity would require a strong restriction of the class of permissible models, and this, in turn, would severely cut into the instrumental powers of simulation.

From the perspective of the ethics of technology, and in the context of ubiquitous computing, Hubig (2008) diagnoses that intransparency is running rampant, and he calls for transparency to be reestablished. From the point of view of ethics and also politics, this is certainly a welcome request that, however, is unlikely to be implemented to any sufficient degree.[10] Definitely, one could and should encourage transparency regarding the specific interests in gathering and using data. However, on a lower and more technical level concerned with models, modules, and how they dynamically interact, a thorough analysis is likely to be much harder, if not impossible. It already takes months to investigate why a computer-supported braking system in a car fails.

Nowadays, one can regularly find examples such as this in the newspapers. Typical cases are making their way from being exceptions to becoming everyday life. Some visionaries were already discussing the growing extension of computer programs back in the 1960s and 1970s. Joseph Weizenbaum, for example, included a chapter on incomprehensible programs in his critical diagnosis (1976). Sheer extension, one could paraphrase Weizenbaum, creates opacity, because changing teams of developers, often over a time frame of several years, cobble together programs whose inner dynamics can finally no longer be comprehended by anybody (Weizenbaum 1976, p. 306).

It seems justified to ask where this expectation (which then gets dashed) comes from in the first place. Weizenbaum conjectures that we have to deal with a suggestion or with an unjustified extrapolation from traditional mathematical modeling and from rational mechanics. The mechanistic metaphor became imprinted as the ideal of explanation

10. Recent critical assessments of big data and big tech stress the point that opacity has become part of the business model; see Foer (2017, p. 73), among others.

during a time when relatively simple machines promoted the Industrial Revolution. Computer modeling is in a way suggestive, because algorithmic mechanisms are executed step by step. However, this does not mean—for reasons of complexity—that the sequence of steps can be followed in practice.

Norbert Wiener already alluded to limiting time as the decisive factor for epistemology: what cannot be understood in an acceptable time cannot be understood at all (Wiener, 1960). This standpoint was further elaborated by Marvin Minsky, who compared a big software program with a complicated network of judicial authorities. Only the interaction of local instances during runtime determines the global outcome, whereas the development of the programs has normally been guided by local reasoning (Minsky, 1967, p. 120).[11]

Let me take stock: Epistemic opacity as a characteristic of simulation modeling is remarkable, above all, because it converts a fundamental feature of mathematical modeling— namely, transparency—into its opposite. This conversion affects practically all epistemologies that take the constructivist property of mathematics as a guarantee for transparency. This applies to Giambattista Vico's "verum et factum convertuntur" ("the true and the made are convertible") that started with the assumption that human beings can understand most perfectly that which they have constructed themselves; in other words, humans and geometry are related like the Creator and the universe. Hobbes takes a similar position. Also Kant, in his *Critique of Pure Reason,* argues that mathematics is the paradigm for a priori knowledge because it creates objectivity via constructions. Nonetheless, it is essential for Kant that constructive activity employs things and relations that are conceptually not fully comprehended.[12]

11. This problem, or set of problems, will be discussed further in chapter 7 in the context of holism and the limits of analysis.
12. It was exactly the lack of conceptual necessity in mathematical constructions that motivated Hegel's critique of Kant, because Hegel saw the mathematical way of knowing as inferior. On the level of the methodology of conceptual analysis, two strategies confront each other here that Robert Brandom aptly called the platonic and the pragmatic. Whereas the former conceives of the use of a concept as determined by its content, the latter takes the complementary direction and starts from how a concept is used. "The content is explicated by the act and not vice versa," as Brandom put it (2000, p. 13). In the case of computer modeling, a platonic strategy seems to be a hopeless approach.

Simulation greatly enlarges the realm of what is feasible, but it pays for this with an increase in epistemic opacity and a loss in analytical understanding. Now, the point is that simulation compensates for this loss—it offers new ways to "understand" model behavior. Exploration, (fast) iteration, and visualization bring researchers, modelers, and model users into close contact with model behavior, or provide access to how the model performs. As a result, orientation in model behavior and options for interventions weaken their dependency on the traditional concept of understanding. Mathematical modeling is able to circumvent epistemic opacity and the complexity barrier—in the sense of finding a surrogate.

Is there a positive side to this? Is it appropriate to introduce a new and weakened notion of intelligibility? At this point, I would like to quote Andy Clark who, in the context of his philosophy of an "extended mind," pointed out that touching, moving, and intervening are essential aspects of human intelligence. From this stance, the kind of analytical understanding mathematics can attain would arguably be an extreme rather than a normal case: "The squishy matter in the skull is great at some things. . . . It is, to put it bluntly, bad at logic and good at Frisbee" (Clark, 2003, p. 5). Simulation modeling, to take a positive look at the matter, taps into exactly these orientation capacities for mathematical modeling.

If epistemic opacity is unavoidable in complex simulation models, and if, nevertheless, possibilities for interventions and predictions exist, will this coexistence lead to a new conception or redefinition of intelligibility? In the closing section of this chapter, I would like to offer a plausible argument that simulation modeling, when it transforms the conception of mathematical modeling, also has the potential for transforming our measures of what counts as intelligible.

INTELLIGIBILITY: AN INSTRUMENT-BASED TRANSFORMATION

The question is whether intelligibility is achievable in complex cases, or whether it is out of reach. Or, is it simulation that puts intelligibility back in reach? I would like to argue for a diagnosis that combines both views: the use of computer simulations leads to a shift in what counts as intelligible; in other words, simulation transforms the meaning of intelligibility.

There are claims stating that simulation offers a new way to attain explanations—namely, explanations based on the interaction of several partial dynamics. The edited volume *Growing Explanations* (Wise, 2004) is devoted to exploring these new options. In the field of social simulations, these are even the standard account. However, in my opinion, one has to face the risk of leaping from the fire into the frying pan. In other words, simulations should not be accepted as providing explanations solely on the basis of their creating emergent phenomena. By itself, this creation has only weak and questionable explanatory power, especially when accompanied by epistemic opacity.

For Humphreys, epistemic opacity points toward the deficient character of simulation-based knowledge. The latter does not meet the high explanatory potential normally ascribed to mathematical theories. From a different perspective, one can also gain a positive side from the simulation method, as has been mentioned, because simulations, notwithstanding their opacity, can open up pragmatic approaches to control and interventions. In such cases, simulations may circumvent the complexity barrier that had prevented theoretical access.

The various examples I have discussed before—numerical optimizers who are using simulations to explore the virtues of certain procedures, climate scientists who vary parameters to find out something about which track hurricanes will follow, or nanoscientists who are looking for ways to reduce friction on the nanoscale—coincide on one point. In all these cases, simulation modeling served the goal of establishing an action-oriented understanding that would be helpful for exerting control and for making predictions, even if the entire model dynamics could not be grasped or comprehended in a theoretical way. In the past, however, scientific understanding used to denote more things and different things than does this action-oriented or pragmatic understanding.

In the philosophy of science, the standard account of scientific understanding derives from explanatory power. Roughly put, if a phenomenon is explained, it is fully understood. Of course, there are major differences regarding what should count as a successful explanation, as shown by the controversy about unificatory and causal explanations (Kitcher & Salmon, 1989). In general, understanding is discussed as being a preliminary or heuristic phase before attaining explanation.

De Regt and Dieks (2005) deserve recognition for having put the concept of understanding back on the agenda of the philosophy of science. They acknowledge explicitly that interpretations of what "understanding" in the sciences means vary considerably—in terms of both the historical evolution of this notion and the differences of opinions at one particular time. Consequently, they are very careful when formulating a general "criterion for understanding phenomena" (CUP): "A phenomenon P can be understood if a theory T of P exists that is intelligible (and meets the usual logical, methodological, and empirical requirements)" (p. 150). And quite consistently, from their standpoint, they add a "criterion for the intelligibility of theories" (CIT): "A scientific theory T is intelligible for scientists (in context C) if they can recognize qualitatively characteristic consequences of T without performing exact calculations" (p. 151).

De Regt and Dieks thus defend two statements: First, they take understanding as a capacity—namely, for realizing typical consequences; and second, they consider intelligible theories as a necessary means for doing this. In my opinion, simulation casts doubt on their second statement.[13]

I have already shown how computer simulations may enable researchers to discern characteristic consequences (within the model), although the model dynamics remain partly opaque. Consequently, an intelligible theory is, at least, not a necessary part of realizing such consequences. The alternative route has been described as circumventing the complexity barrier.

Apart from the type of orientation in the model and the approach taken in the process of modeling, which explanatory capacity can one

13. Compare the three contributions discussing the role of models relative to theories written by Morrison, by Knuuttila and Merz, and by Lenhard in de Regt, Leonelli, and Eigner (2009).

ascribe to simulation? On the one hand, the traditional standard of scientific explanation is not met. One who is unable to explicate *why* a certain behavior happens cannot explain this behavior. Moreover, one who has no clue regarding on which grounds a generalization could possibly proceed, and mere iteration does not count here, also has nothing in hand on which an explanation could be based. On the other hand, one has understood a great deal when one is able to create nanowires or make a promising proposal regarding how difficulties with friction on the nanoscale might be overcome.

Hence, one faces two possibilities: either stick to the established notion of understanding and accept simulation as an option for control *without* understanding; or specify a pragmatic account of understanding oriented toward the capacity for prediction and intervention. Then, intelligibility and understanding are present whenever this capacity is present.

The second option is not as far-fetched as it might seem. The meaning of intelligibility in the sciences has always been tied to the instruments available at a certain time and place. What counts as opaque, therefore, is not fixed. In addition, the meaning of intelligibility and analytical understanding depends on the mathematical instruments available. Peter Dear, in his enlightening book *The Intelligibility of Nature: How Science Makes Sense of the World* (2006), has underlined the close interdependence of modern science and intelligibility. I would like to briefly discuss the choice between the two possibilities in a reading of Dear.

Peter Dear juxtaposes two attempts: science as natural philosophy, whose goal is explanation and understanding; and "science as an operational, or instrumental, set of techniques used to do things: in short, science as a form of engineering, whether that engineering be mechanical, genetic, computational, or any other sort of practical intervention in the world" (Dear, 2006, p. 2). Dear concedes that, ever since the seventeenth century, both directions have not existed as single and pure movements. Rather, they are two ideal types in the sense of Max Weber. I would like to add that both lines of thought are not only present in science but also influence each other.

The most interesting point, in my view, is the perspective on Newton. For Dear, it is clear that Newton failed as a natural philosopher but not as a mathematician. His famous verdict *hypotheses non fingo* in the *Scholium* of the *Principia* expresses that Newton, despite all his efforts, did not feel able to specify the mechanisms of gravitation that would be required to fulfill the (then) commonly accepted standards of intelligibility. Regarding mathematization, of course, his theory is one of the greatest-ever achievements in science.

The philosophically and historically significant twist is that the criteria for intelligibility are variable. In fact, these criteria are exactly adapted to Newton's theory. "The result was that action at a distance eventually came to be seen, almost by default, as an explanatory approach that was satisfying to the understanding" (Dear, 2006, p. 13). In my view, this is an equally convincing and significant instance of the power of mathematical instrumentation—namely, influencing what counts as intelligible.[14]

Dear stresses further that instrumentality alone is not the single factor driving the development of the sciences. It is, rather, intelligibility—that is, the tradition of natural philosophy—that takes a central role (Dear, 2006, p. 175). Dear rightly insists on this role, but he misses, in my view, the somewhat dialectical point that intelligibility and instrumentation are interdependent. Where Dear honorably defends the significance of intelligibility in the face of the strong ideology of instrumental feasibility during the twentieth century, a supporter of option two, without contradicting him directly, sees things as being more fluid. The redefinition of intelligibility is a facet of the changes that proceed from the transformation of mathematical modeling I am currently investigating. Intelligibility is about to change again owing to the use of the computer for modeling.[15] This does *not* mean, however, that a lack of insight can no longer be criticized because it has simply been defined away.

14. Dear discusses further examples: Maxwell and Faraday, as well as Darwin versus Herschel.

15. Gunderson considers this option—related to the meaning of "reasoning"—in his classic paper on artificial intelligence: "with the increasing role of machines in society the word 'think' itself might take on new meanings, and that it is not unreasonable to suppose it changing in meaning in such a way that fifty years hence a machine which could play the imitation game would in ordinary parlance be called a machine which could think" (1985b, p. 58).

To what extent and in which ways is the meaning of understanding actually undergoing a change? Simulation-based understanding is not merely a surrogate for analytical understanding. Even if one were to accept that the latter provides a gold standard, this standard is not universally applicable. Investigations of applied and engineering sciences are increasingly revealing the limitations of this standard. Not only would this standard limit scientific practices, it would also narrow philosophical investigations. Norton Wise has expressed this opinion neatly:

> I have come increasingly to think that the pretty mathematical theory (let's say quantum mechanics) actually presents the lower standard because it explains only idealized systems and cannot deal with much of anything in detail in the real world, unless replaced by a simulation. The idealized theory may very well help in understanding the simulation but the simulation often goes much further. (personal communication, Jan 2015)

Spelling out the ways in which simulation goes "much further" is tantamount to answering the question of how understanding is transformed. The present chapter has tackled this question. Nonetheless, it has been more successful in showing the relevance and timeliness of the question rather than in reaching a full answer. The pragmatic understanding described here might be seen as a first step toward theoretical understanding, as Parker (2014a) argues. Tapping resources from model-based reasoning seems to offer a way forward, as MacLeod and Nersessian (2015) show. Formulating an answer is, in my eyes, one of the most important challenges facing the philosophy of simulation.

5

A New Type of Mathematical Modeling

The phenomena investigated in chapters 1 through 4 correlate positively; that is, they interact in such a way that one strengthens the other. As a result, they converge to form a new type of mathematical modeling. Simulation modeling is distinguished by its exploratory and iterative mode that presents a multidimensional picture. The properties spanning the dimensions can occur to different degrees; that is, to what extent the characteristic properties are realized depends on the particular classes of simulation strategies and the concrete applications. The decisive point, however, is that these characteristic properties are linked and reinforce each other. From a systematic perspective, I shall argue, simulation synthesizes theoretical and technological elements. Hence, simulation-based sciences are application oriented and show a greater proximity to engineering.

In what respect, if any, is simulation a novelty in the sciences? The present chapter represents my first attempt to answer this question. My vote is akin to the basic claim underlying this entire investigation—namely, simulation is a new type of mathematical modeling. In the introduction, I mentioned that any diagnosis of what is novel depends on what categories or criteria are chosen. On a small scale, almost every scientific approach entails a couple of novelties; whereas on a large scale, it might seem as if hardly anything new happens at all. For a philosophical investigation, it is important to identify the right scale or the adequate degree of abstraction. Here, I want to defend a moderate position that first classifies simulation modeling as a type of mathematical modeling and then specifies essential traits of simulation that justify speaking of a *new* type of modeling.

I shall qualify this answer in a second approach, arguing how simulation modeling transforms fundamental notions such as those of solution (chapter 6) and validation (chapter 7), as well as the dichotomy between realism and instrumentalism (chapter 8). In sum, this elevates simulation modeling from a type of mathematical modeling to a new combinatorial style of reasoning (chapter 8).

MULTIDIMENSIONALITY

The features of simulation modeling examined in the previous chapters occur in different degrees. The single elements are

- **Experimenting.** Simulation experiments are a special class of experiments. They are particularly relevant for simulation *modeling* and especially for the exploratory variant of experimentation. Model assumptions with effects that are hard or even impossible to survey can be tested, varied, and modified by applying iterative–exploratory procedures. Modeling and experimenting agree to engage in an exploratory cooperation. Such cooperation regularly employs artificial elements.
- **Artificial elements.** These are unavoidable in discrete modeling, given that standards of accuracy play a role. Either a theoretical model is formulated in the language of continuous mathematics and has to be discretized, or one is dealing with a discrete model from the start. When controlling the performance of discrete models (i.e., for instrumentalist—though unavoidable—reasons), artificial components are included. Experiments are necessary to adapt the dynamics of a simulation model to the phenomena to be modeled, because one cannot judge whether these artificial elements are adequate without such experimental loops. This grants simulation modeling an instrumental aspect that blurs the representation relation and hence weakens the explanatory power.

- **Visualization**. The computational capacity has to meet high demands from two sides: either from the side of the data that, in many applications, have to be handled in great amounts; or from the internal side of the model itself that might require extensive experiments and iterations during the construction phase. Interfaces between modelers and computers are indispensable for guiding the further process of modeling. Visualizations offer ways to present such huge masses of data. Of special interest for the process of modeling are those visualizations that help to establish interactions between modelers, models (under construction), and the dynamics they generate.
- **Plasticity**. This denotes the high level of adaptability in a simulation model's dynamics. The structural core of such a model is often no more than a schema that requires—and allows—further specification before simulating particular patterns and phenomena. Whereas theoretical considerations enter mainly into the structural core, specification follows an exploratory approach. However, both structure *and* specification are necessary to determine the dynamic properties of a model. It is this contribution of specification that renders simulation models structurally underdetermined.
- **Epistemic opacity**. This arises because models are becoming more complex in several respects. The course of dynamic events encompasses an enormous number of steps, so that the overall result cannot be derived from the structure. Instead, it emerges from model assumptions and the parameter assignments chosen during runtime. Additionally, important properties of the dynamics result from the specifications and adaptations made while developing the model. This reveals a fundamental difference compared to the traditional concept of mathematical modeling and its concern with epistemic transparency. Even if simulation cannot reestablish the epistemic transparency of classical mathematical models, it might nevertheless be able to provide options for control and intervention by utilizing experiments, tests, controlled variation, and so forth.

These features are interdependent, and this is what turns them from a set of features into a distinct type with components that systematically reinforce each other. Simulation modeling follows an exploratory and iterative process that involves the aforementioned components—either utilizing them or compensating for them (opacity). This procedure not only reduces but also *channels* complexity. Any computer-based modeling has to respect the limitations of computational complexity. For instance, it is often the case that even the computer cannot consider all the possibilities, for reasons of complexity. At the same time, iterations can be performed in extremely great numbers and at high speed; and this, in turn, supports exploration. Hence, modeling can proceed via an iterated feedback loop to performance criteria.

The result is a multidimensional picture of simulation modeling. The various simulation techniques and illustrative examples are distinguished to different degrees along these dimensions. There are extreme cases in which just one single dimension is practically relevant. If, for instance, a mathematically defined probability distribution is simulated in a Monte Carlo approach, then arguably the only component at work is the experimental one. All the other dimensions, however, make sense only when viewed together. Employing artificial elements will regularly generate opacity, and can therefore be coordinated only in an iterative–exploratory way. Likewise, the plasticity of models requires an exploratory approach, because the necessary specification cannot be found in any other way in lieu of structural criteria. Visualization, finally, makes model behavior accessible to modelers. Hence, it helps to make the modeling process feasible in practice.

In summary, the components of simulation modeling do not merely occur together coincidentally but, rather, depend on each other and fit together to form a distinct type in which epistemological and methodological aspects are intertwined. I have shown this result for a wide array of model classes, including finite differences, cellular automata, and neural networks. Because the components are interdependent, it would be misleading to highlight just one single dimension as being *the* relevant one for simulation—be it the experimental character, the new syntax, or the epistemic opacity.

NOVELTY?

When complexity is a special feature of simulation, one could suspect that mathematical modeling follows a more general trend here that can be attributed to all the sciences.

According to this viewpoint, societal and political expectations and demands do not leave any other choice for science than to address complex application contexts. On the one hand, it seems to be unacceptable to reduce complexity by simplifying and idealizing, because this would sacrifice the relevance of the results. On the other hand, following this rule exhausts the transparency reservoir of the sciences. Increasing complexity poses a dilemma, and a growing number of publications are analyzing this diagnosis from different perspectives.[1] Simulation shows a certain affinity with complexity, insofar as the modeling *itself* normally generates complexity—while also calculating with it. Under the conditions of (computer) technology, the mathematical models themselves become complex, so that simulation modeling is not simply an extension of the process of mathematization.

During the pioneering phase, and also during the ensuing decades in which computer models slowly became more established, their use remained somewhat exotic. Compared with theoretical approaches, computer modeling was seen as a method not of first but only of second choice.[2] Up to the 1980s, simulation had not yet become customary. A typical example is Robert Brennan's essay *Simulation Is Wha-a-t?* in McLeod's edited book on simulation (1968). It was only after computer methods had become widely adopted by the sciences that the *philosophy* of science started to raise questions about the status and novelty of simulation and computer modeling. The session at the 1990 PSA conference, organized by Humphreys and Rohrlich, provides an early instance. Since then, at regular intervals, the discussion has started anew about whether simulation is novel, and if it is, what actually characterizes its novelty. This discussion is developing between two main opposing camps that

1. This includes works such as Gibbons et al. (1994); Weingart, Carrier, and Krohn (2007); or Mitchell (2009).

2. For a periodization of computer modeling, see Lenhard (2017c).

can be called—with a grain of salt—serene conservatives and emphatic prophets.

The first camp of serene conservatives favors a stance that is not unknown in the history of philosophy—namely, *nil sub sole novum*. According to this camp, simulation is nothing remarkable in a philosophical respect. Simulation is a new instrument, but its novelty consists merely in the increased speed of computational processes, whereas in a philosophically more fundamental respect, simulation has little to offer. Stöckler (2000) or Frigg and Reiss (2009) articulate this viewpoint.

Of course, if simulations are understood merely as tools for numerically approximating an already settled and fixed mathematical truth, then simulation models lose any autonomous or semi-autonomous properties. According to such a position, simulations remain a conceptually subordinate, merely auxiliary tool. The argumentation in the previous chapters and, independently, the broader philosophical debate over models, have refuted this position. Somebody who nevertheless adheres to such a view will admittedly not need to undertake a thorough investigation, but will be convinced that simulations will not offer anything novel because they are model based and of a subordinate status.

In my view, there is at most only partial justification for this position. It is justified to a certain extent, because important components of simulation modeling such as experimenting or using artificial elements are well established in the sciences. These components (ingredients) do not emerge from (a precomputer) void. However, one can criticize this position, because it throws out the baby with the bathwater. Even if no particular ingredient is of tremendous novelty, the way these ingredients combine may well be novel. The passage about gunpowder cited in the book's introduction illustrates that fact. The argumentation in this book aims exactly to bring out the characteristics of simulation modeling *within* (or against the background of) the framework of mathematical modeling.

In contrast, the second camp of scholars, the emphatic prophets, highlights the novelty of simulation. Keller (2003) expresses this emphasis as follows:

> Just as with all the other ways in which the computer has changed and continues to change our lives, so too, in the use of computer

> simulations—and probably in the very meaning of science—we can but dimly see, and certainly only begin to describe, the ways in which exploitation of and growing reliance on these opportunities changes our experience, our sciences, our very minds. (p. 201)

Galison (1996, 1997) is another prominent representative of this camp. In my view, such claims are productive and worth pursuing, because only a critical investigation can explore the limits of such theses. Simulation will find its true status through a critical comparison of both positions.

Here, I would like to uphold an intermediate—or perhaps even mediating—position. In philosophy, novelty is rarely novelty per se, but is very likely to be novelty from a certain perspective. The present investigation presents simulation modeling as a *new type* of mathematical modeling. Hence, it stresses novelty as much as it subsumes the issue under the category of modeling. In this respect, I regard the stance that Humphreys laid down in 2004 and accentuated further in 2009 as being closely allied: the analysis of simulation gains in contrast and depth when placed in the wider context of philosophy of science—namely, the debate about modeling and, more specifically, about the mediating and autonomous role of models. The close (and novel) interdependence of technological and mathematical aspects in simulation modeling amplifies the autonomy of simulation models.

The internal aspects of simulation modeling that concern its peculiar dynamics take up ample room in the activities of modeling. This intensifies the autonomous side of the mediating role; see also the final passages of chapter 1. In this respect, simulation does indeed present an intensification triggered by computer technology. In the present context, autonomy is always meant to be partial; otherwise, it would hardly be possible to increase it. I use autonomy in the same sense as Morrison (1999), who introduced it in the context of modeling. The increase is based on a significant shift compared to the traditional, analytically oriented conception of mathematical modeling. It shows the strong influence of application-oriented modeling with its tendency to mix the roles of models, data, and semi-empirical, exploratory procedures.[3]

3. Apt analyses of the relationships between theory, data, and models in applied fields can be found in, for instance, Carrier (2010) or Wilholt (2006).

In my perception, one advantage of the exploratory type of modeling is that it combines two motives that are both well established in the sciences but apparently oppose each other. Mathematical modeling ties in with the high degree of lasting and certain knowledge that is the promise of mathematics. At the same time, the exploratory mode of simulation stresses its open and tentative character.

In this respect, the characterization of simulation I am developing here is akin to what scholars such as Hans-Jörg Rheinberger intend whose works on "experimental systems and epistemic things" start from the "primacy of scientific experience that is still developing" (2006, p. 27; my translation). Rheinberger also speaks about the provisional status of epistemic things (p. 28) and links up here with older positions such as those of Claude Bernard or Ludwik Fleck. These considerations fit nicely with the exploratory mode of computer simulation, because this mode grants a strong influence to the theoretical side without allowing the openness of the further course of modeling to disappear. Simulation modeling might achieve a mediating balance here—at least in successful cases.

When artificial elements are employed and tested, or when parameters are varied, the resulting performance (if it succeeds) is definitely not owed to blind but, rather, to systematic trial. Being systematic here encompasses reference to theories. When (pseudo)random experiments are carried out, as in Monte Carlo methods, variations are random in what can be called a systematic way. The other pole is contingent variation according to knowledge of how the model behaves. In both cases, preliminary results occur, but these are preliminary within the framework of a systematic and controlled procedure. Put briefly, simulation modeling presents a distinct type of mathematical modeling. Notwithstanding its continuity with notions and procedures of traditional mathematical modeling, this type also includes essential differences.

CONVERGENCE WITH ENGINEERING

It is obvious that simulation methods build links between the natural sciences, a part of the social and economic sciences, and the engineering sciences. In all these sciences, simulation methods drive a similar kind of mathematization. However, mathematization does not mean a process of abstraction and formalization (although it might mean this). Instead, simulation methods are a conglomerate comprising mathematical, computational, institutional, and infrastructural elements.[4] For instance, new probabilistic methods such as Markov chain Monte Carlo (MCMC) thrive in many fields and are tailor-made for complicated models that handle vast amounts of data. These methods are employed in practical work while also constituting an area of active research in mathematics. There, researchers struggle to theoretically justify methods and procedures already established in practice (see Diaconis, 2009). Software languages build another interesting aspect of how this conglomerate develops. It was only when semantically advanced software languages became available—such as object-oriented programming and C++—that it first became possible for scientists to follow an exploratory mode. Such a mode is making computational modeling amenable to many scientific problems, but this presupposes that frequent modifications in code are easy. Before the advent of advanced languages, coding required technically skilled experts who often focused exclusively on the copious ways of programming at "machine level."

Developments on the hardware side are of equal importance. Visualizations, for instance, that are easy to produce and integrate are equally essential for both engineering and the natural sciences. Finally, on the side of infrastructure and organization, networking plays a major role. Networked infrastructure, today often taken for granted, is a relatively recent phenomenon. Without it, it would hardly be possible to exchange software libraries and entire submodels or to tackle comprehensive modeling tasks by distributing single subtasks. All these aspects are signs of a convergence of science and technology—a convergence that thrives on comparable, surely not identical, canalizing effects in all scientific fields that employ simulations.

4. The Boston Studies *Mathematics as a Tool* (Lenhard & Carrier, 2017) assembles philosophical and historical analyses of tool-oriented mathematization.

In this respect, modeling has to respect technological conditions, as well. This is clear to see, because simulations inevitably have to be implemented effectively on a machine. They have to "run" if they are to serve as simulations. In particular, the philosophical focus switches from considerations *in principle*, such as the solvability of equations, to considerations *in practice* (as Humphreys pointed out in 2004), such as finding usable results with acceptable time and machine resources.

Beyond that, the convergence thesis is supported by the mode of modeling itself: when employing simulation methods, natural and engineering sciences converge.[5] This is a significant observation that I shall take up again in chapter 8. Simulation seems to be tailor-made for the needs of technology or of those sciences that are oriented toward technology. The decisive point for technological applications is not reference to universal laws but, rather, grip in concrete situations. What are the phenomena of interest, what are the relevant quantitative measures, and what concrete prediction follows? What matters are the particular initial and boundary conditions—of materials, the context, the temporal order, and so forth. Only knowledge of this particular kind can be applied successfully.

Hence, the relevance of knowledge about laws is somewhat restricted, especially when such knowledge founders if it comes up against the complexity barrier—that is, if it does not suffice to determine model behavior.[6] What is actually needed is recipes in the sense of design rules. This lowers the demands for theoretical insight. Nevertheless, a certain degree of generality is necessary even then, because design rules do not work like an instruction manual for the defined uses of one particular device. Instead, to be useful at all, a rule has to be resilient and—to a certain degree—applicable in a variety of circumstances. The conditions relevant in technological applications lie somewhere between general theoretical knowledge and special instructions.

5. Moreover, the engineering sciences are influenced by computers and simulation. The status of *computational sciences*, as well as material sciences, would then be in need of further explanation (cf. Bensaude-Vincent, 2001; Johnson, 2009).

6. The classic reference is Nancy Cartwright's influential *How the Laws of Physics Lie* (1983). I have already discussed the complexity barrier in connection with simulation in chapter 4.

One could point out initially that theoretical knowledge of laws would be the best starting point for deriving useful recipes—nothing is more practical than a good theory. In general, however, this is not quite the case. For instance, nuclear fusion is theoretically well understood, but there still remains a leap to being able to actually build a reactor, because a concrete realization poses its own complex problems (cf. Carrier & Stöltzner, 2007). In the case of the golden nanowire forming between the surface of the gold and the nickel tip, simulation experiments exhibited surprising behavior. Although the phenomenon was confirmed later, the simulation did not "explain" it in the usual sense. Of course, law-based knowledge was implemented in the simulation model, but the relationship between the Schrödinger equation and the golden wire remained opaque. Landman was certainly surprised when observing a wire, but he could not gain insight into why such a wire emerged. Although the simulation obviously employed theory, it did not offer theory-based insight. What it did offer instead was understanding in a pragmatic sense, tied to the potential to intervene and control, and therefore showed a great affinity to technological artifacts. This kind of pragmatic understanding is an important factor for the convergence of simulation-based sciences, and I would like to highlight it as a novel factor in the epistemology of mathematical modeling.

The close relation between science and technology continues in testing and validation. Testing technological artifacts and controlling model behavior are analogous and differ from situations that are theoretically transparent. Iterative and exploratory modeling can search for possibilities but not to an arbitrary extent. Such a procedure is similar to probing devices on a test bed. Chapter 7 will discuss the problems of validation in more detail.

The partial independence of engineering knowledge from theoretical scientific knowledge has been stressed in the philosophy of technology (see, e.g., Vincenti, 1990). Admittedly, technological innovations might result from scientific research, but efficient design rules do not need to piggyback on theoretically mastering the matter under consideration. Partial autonomy, however, does not equal full autonomy.[7]

7. Wilholt (2006) provides an instructive example in which design rules work by mediating among phenomena, data, practical goals, and theories.

At this point one must be careful. The analysis of simulation modeling has shown how science and technology are closely interrelated. It would be premature, however, to attribute this interrelationship to simulation and the use of computers. In fact, science and technology were intermeshed long before computers existed. Simulation gives only a reason to take a new and focused look. My argumentation about artificial, nonrepresenting components, for instance, is supported by Scerri's (2004) work showing that *ab initio* methods of quantum chemistry actually do integrate semi-empirical components. At the same time, other philosophical studies indicate that semi-empirical approaches are widespread in the sciences (Ramsey, 1997). There is a widely held view that sciences are set strictly apart from engineering disciplines and technologies. In my opinion, this view is biased because it ignores applied practices.[8]

My opinion corresponds to a series of philosophical and historical standpoints without agreeing with them in every detail. In the history of technology, there is a growing community arguing that the perspective on the history of the sciences has been seriously hampered by systematically neglecting applied sciences. Derek deSolla Price followed a similar direction when he stated: "The lack of critical scholarship on the interaction between science and technology is due to the separation of history of science and history of technology" (Price, 1984, p. 105).

Hence, I am particularly interested in those alternative theoretical approaches that specify how science and technology are related. Be it a symbiotic amalgamation, as in the concept of "technoscience" à la Latour or Haraway; or a mutual influence of fundamental science, applied science, and technology, as proposed by Carrier (2010) in his concept of application-dominated science, I prefer a standpoint that leaves room for interaction between science and technology. Simulation modeling, I maintain, occurs in a mode that intensifies this interaction. Any position merging both positions, or reducing one to the other, could not take intensification seriously. "That is, postmodernity does not reduce science to technology but recognizes their intimate interrelation. This is a cultural/scientific shift that I very much welcome. It is opening up

8. Radder (2009) discusses perspectives on this divide.

new kinds of knowledge and knowledge of new kinds of things" (Wise, 2007, p. 6). And when Wise states further that the pursuit of knowledge is much more closely tied up with pursuit of technologies than ever before, his point gains particular support from simulation.[9] Such a view coincides with an understanding of "technoscience" as defended by Nordmann (2010). In his analysis, the concept involves the close entanglement of both components, but does not suppose any reductionist relationship.

Like all artifacts, simulation models develop their own dynamics that bestow upon these models a semi-autonomous character. Some modeling steps that arise from the model's own life appear as errors or falsities under a certain and overly simplistic perspective: a perspective according to which mathematical constructs (in particular, the discrete computational models) are referring to a completely determined external reality, a reality moreover that is assumed to be contained in theories formulated by continuum mathematics. It is only under these presuppositions that artificial components appear to be "false." Margret Morrison and many colleagues have highlighted the autonomy of models (see the discussion in chapter 1). This insight applies all the more to simulation models.

In the preceding argumentation, I advocated a philosophy of science and technology that is sensitive to the properties of instrumentation.[10] A philosophical investigation of simulation and computational modeling has to take computers, models, and theories into account *together*. One might look at scientific theories and computers as two opposing, or independent, elements, but this perspective would quickly lead into uninteresting territory, which would no longer offer "novelty."

9. Wise supports his claim by referring to computer simulation, as well as model organisms. I also see certain similarities between the two (Lenhard, 2006).

10. According to Don Ihde (2004), this standpoint characterizes the philosophy of technology.

PART II

CONCEPTUAL TRANSFORMATIONS

6

Solution or Imitation?

Simulations are often described as numerical solutions, and this description is equally widespread in both the specialist user literature and the philosophy of science. It suggests that the notion of solution remains invariant, a fixed point for characterizing simulation; in other words, it is simply that one and the same equations are being solved numerically instead of analytically. The description of simulation modeling in the preceding chapters, however, brought to light an interesting complication. Numerical solutions, adequately analyzed, are a quite different concept of solution from analytical solutions. Because the process of modeling involves a number of transformations, the properties of a certain numerical solution will differ regularly from the properties of the (often unknown) analytical solution. The latter is defined by the mathematical rules of calculus. Numerical solutions, however, are a multifaceted affair encompassing efficiency, adequacy, tractability, and usability. The present chapter presents a critical discussion of how the traditional notion of solution and the new, simulation-related notion of numerical solution differ.

Recall chapter 1, in which I looked at the simulation modeling of the atmospheric circulation. There, theoretical mathematical models took a main role. These models captured the atmospheric dynamics in the form of a system of partial differential equations. These equations, in turn, were remodeled to transform them into finite difference equations. In the end, it became clear that the simulation model does not solve the original system of equations. Rather, it simulates the atmospheric dynamics in an adequate and, most of all, numerically stable way.

When talking about a solution in the proper, strict sense, what is meant is the mathematical notion of solution. In the example just mentioned, this is the mathematical term that satisfies the system of differential equations. This strict notion is relevant in a broad class of natural and engineering sciences—namely, those employing the language of differential and integral calculus or, even more generally, those employing the language of mathematical equations to formulate theoretical models. Simulation is often conceived of as numerical solution (see, e.g., Dowling, 1999; Hartmann, 1996), but, as I shall show here, this manner of speaking requires a more differentiated approach.

What counts as a solution in the strict sense depends solely on the mathematical equations to be solved (including definitions of the underlying mathematical space and the like). A quadratic equation, for example, can be written as

$$a \cdot x^2 + b \cdot x + c = 0$$

and solutions are exactly those numbers x that satisfy this equation (for the given numbers a, b, and c). Though there might exist several and, in particular two, solutions, the criterion is strict: if you insert the proposed solution, it will either satisfy the equation or it will not.

In the present case, there is even an exact formula that delivers the solution. However, when cases become more complicated, such a formula no longer exists. Instead, there might well be numerical procedures for finding a proxy to the solution that are practicable even without a computer. Algorithmic procedures for determining a local minimum provide an example. However, if you start with a continuous model and then replace it with a discrete simulation model, then in general the solution for the simulation model will not solve the continuous one. This would require an exact match between the two models.

Alongside the *strict* sense of solution there is a *pragmatic* sense in which you can speak of solving a problem numerically. Such a solution would then be good enough to solve the problem at hand in a numerical–quantitative sense. For instance, you could search for a number that satisfies an equation up to a certain level of inaccuracy—that is, one that if inserted into the equation, would produce a value close to zero. If the criterion for what counts as being close to zero is specified precisely, you have determined what does and what does not count as a solution. In this case, the solution is defined via a condition of adequacy that can be, but need not be, mathematically precise. Such a condition could well depend on what works well enough in a certain context. Hence, the pragmatic sense of solution is logically more complex than the strict one. In the meteorological example given in chapter 1, the simulation provides a pragmatic (but not a strict) solution of the dynamics of the atmosphere. The simulation generates fluid dynamic patterns that are similar enough to the observed ones, although what counts as adequate is not defined in a quantitative and precise way.

The pragmatic concept of solution is obviously more general than the strict concept in which the context is, so to speak, already fixed completely in the equation. The pragmatic concept allows a much wider scope of application and does not necessarily follow a mathematically strict criterion. Also, it is complicated in an interesting sense because of the need to balance aspects of efficiency, economy, and adequate accuracy. Nothing like this task is to be found in the traditional notion of a mathematical solution.

Computer simulations are always numerical. If successful, they provide a pragmatic solution and maybe even a strict solution, but this would be an exceptional case.[1] The strict concept of solution is in fact more special:

1. It is only in cases in which a theoretical and mathematically formulated model serves as a basis that a strict solution is defined at all.
2. Even if a strict solution is defined, simulation modeling will hardly ever really deliver it.

Identifying simulation as a numerical solution has to be handled with care, because it mixes up the strict and the pragmatic sense of a solution. Speaking in this way supports a commonly supposed but misleading

1. I refer readers interested in the intricacies and conceptual richness of numerical solutions to the thoughtful accounts in Fillion and Corless (2014).

picture in which simulation is a straightforward continuation of traditional mathematical modeling. Simulation allegedly would solve numerically those problems that cannot be solved analytically—comparable with a motor vehicle that gets stuck in the mud, but then can continue moving when switched to four-wheel drive. However, a closer look at simulation modeling reveals that it is not similar to this metaphor. A more adequate picture has to take into account those modeling steps to be found only in simulation modeling: those that combine theoretical with instrumental components. Therefore, I maintain that the simulation model (only) *imitates,* rather than solves, the dynamics of the theoretical model. This thesis has been presented in chapter 1 for the case of finite differences and in chapter 2 for the case of cellular automata. It can be useful to look at simulations as numerical solutions. Then one has to be aware of the fact that numerical solutions imitate ordinary solutions and that the quality of imitation depends on various pragmatic criteria. This makes numerical solutions conceptually different from traditional (strict, analytical) solutions.

In the following, I want to distinguish two fundamental but opposing conceptions of simulation. As I shall show, neither position can be defended in its pure form. The first conception starts from the observation that analytic means, in the technical mathematical sense, are often insufficient to solve mathematical equations. Consequently, simulations are conceived as *numerical solutions* of these equations. The second approach follows a different line of thought that does not involve the concept of solution. It takes simulation as the *imitation of the behavior* of a complex system by a computer model.

These two conceptions conflict, and this conflict has shaped the typology of simulation right from the start. That is something I shall illustrate by analyzing the relationship between the simulation pioneers John von Neumann, who advocated the solution, and Norbert Wiener, who advocated the imitation concept. However, to start with, both conceptions of simulation—solution and imitation—rarely occur in their pure form in practice. Indeed, as simulation modeling consolidated and spread out to many fields of the sciences, it took a form that *combined* both types. Nonetheless, the relative portions of solution and imitation might vary from case to case—as a couple of examples will show.

The thesis of combination is based on the insight that simulation modeling has to mediate between two conflicting forces: theoretical understanding and epistemic quality stand on the one side; applicability and tractability is on the other. Large parts of the sciences involve a compromise (in one way or another) between these two diverging forces. What is interesting about simulation is the way in which a balance is achieved—that is, how the conflicting types are combined.

TYPOLOGY AND PIONEERING HISTORY

Essential parts of the pioneering work in developing electronic computers were carried out during the 1940s. This was mainly government-funded research, particularly for the military. The history of computer development is a multifaceted story in which engineers, scientists, technologies, and institutions, as well as mathematical and logical concepts, all have their roles to play.[2] I would like to pick out John von Neumann and Norbert Wiener, two scientists whose relationship shows something exemplary about simulation. Both were brilliant mathematicians who were involved in military research during World War II. Heims (1980) has devoted a valuable monograph to juxtaposing the two persons, in which he discusses how both scientists directed their research and ambitions in different ways. I would like to add something to Heims's investigation—namely, the following philosophical thesis: the controversy between von Neumann and Wiener did not rest merely upon their idiosyncratic relationship but also upon their complementary conceptions of simulation modeling.

Initially, both men had the vision that the new technology of the computer would present a great opportunity for new mathematical modeling approaches. Both planned to combine their powers with colleagues and form an interdisciplinary group. The group would explore the new

2. Ceruzzi (2003) offers a historical presentation oriented toward machines. The monographs of Edwards (1996) and Akera (2006), written from a science studies perspective, instructively embed the history in the context of the Cold War.

opportunities that the computer, still in its developmental stage, would open up when used as a tool for the sciences. This context saw the launch of the Teleological Society, later called the Cybernetics Group. The first meeting, at Princeton in January 1945, was secret because the group had to circumvent regulations on research of potential military interest. From 1946 onward, they met in New York, where they organized a total of ten conferences at the Edith Macy Conference Center. However, the collaboration between the two men did not last long, because von Neumann soon left the group.

At that time, Wiener, who worked at MIT and had cultivated a passion for engineering sciences, was mainly concerned with the design and construction of a concrete device. He was collaborating with the engineer John Bigelow on behalf of the National Defense Research Committee (NDRC). Their goal was to develop a computer-based guiding system for an antiaircraft gun, called the Anti-Aircraft Predictor. With good reasons, Galison (1994) and Heims (1980) claim that cybernetics emerged from this project.

This work made direct use of Wiener's mathematical–statistical theories about filtering data, a very general account of prediction under incomplete information. Independent of this but at the same time, Wiener and Claude Shannon were formulating the theory of information from a radically statistical perspective. As is well known, the amount of information in this technical sense is something completely different from what common sense takes as content or meaning. No regular utterance has maximal information. Even if some syllables or words have been misunderstood, one can normally understand the message. That means one can reconstruct the missing parts. The amount of information (in the technical sense) measures the degree of unpredictability. Some part of a message has maximal information only if it cannot be reconstructed (inferred) from the rest of the message. In this sense, a purely random sequence of numbers is noise for human beings, but every number contains maximal information, insofar as knowledge about all numbers that precede a certain entry does not help in the least when it comes to predicting this next entry. In other words, this next number has full information content again because all information available up to this point does not help when it comes to guessing this (upcoming) number.

Wiener's project at the NDRC consisted in using mathematical–statistical methods to predict the position of an aircraft on the basis of a few observations of its trajectory, so that the gun could be adjusted accordingly. Wiener ascribed philosophical significance to this military research and development (R&D) project. He conceived of it in a broader context. For Wiener, the particular task of the project matched the fundamental epistemological situation of human beings who are characterized by ignorance. Under such a condition, the general goal is to act purposefully or rationally on the basis of incomplete information. Statistical pattern recognition for him was the modeling strategy that dispensed with detailed assumptions about the mechanisms that had produced the patterns. In this way, one could compute with principal ignorance, and compensate for this through mathematical instruments that would make predictions possible. In the case of the antiaircraft predictor, this definitely had the goal of carrying out an intervention.

In the more concrete terms of the NDRC project, this meant predicting the position of an aircraft based on minimal assumptions about the behavior of the pilot. Thus, prediction does not mean predicting novel phenomena but, rather, predicting in the logical sense of continuing a pattern in which the first parts are known. Wiener was fascinated by the military task of not only determining the position with theoretical methods but also of constructing a concrete device including servo-electrical actuators—the prototype of an automated guiding or pointing system. The project, however, was never completed. Nevertheless, it started Wiener's long-lasting interest in control systems and teleological behavior. He ascribed high philosophical significance to the strong connection between the two. When the new technologies of automated control converged with mathematical statistical theories, Wiener thought, this would be the dawn of a new epoch, the cybernetic era:[3] "If the seventeenth and early eighteenth centuries are the age of clocks, and the later eighteenth and nineteenth

3. See Wiener (1948). He saw himself as the principal founder. Mindell (2002), however, argues that this was a myth promoted by Wiener himself, who overestimated his mathematical achievements and did not take adequate account of the context and significance of the engineering sciences.

centuries constitute the age of steam engines, the present time is the age of communication and control" (Wiener 1948, p. 39).

John von Neumann also realized the significance of new computing technology at an early stage. In Göttingen, Germany, he had worked on the mathematical foundations of quantum mechanics before immigrating to the United States, where he joined the Institute for Advanced Studies (IAS) in Princeton, quickly becoming one of the most influential mathematicians of his time. He was involved in the development of the conceptions of digital technology[4] and also played a role in the invention of simulation approaches, as Galison (1996) describes for the case of the Monte Carlo simulations created in the context of the Manhattan Project. The theory of automata would be another case that von Neumann advanced, based also on helpful ideas from Stanislaw Ulam, such as the theory of cellular automata.[5] In brief, both protagonists were asking themselves what the new scientific discipline would have to look like that was going to raise the computer to the central instrument, and what type of mathematical modeling would be appropriate for this discipline. Moreover, both started from mathematics and inquired about what computer-based "solutions" would have to look like.

Despite all their common ground, von Neumann advocated a much more formal and, if you like, a more optimistic standpoint. He saw the computer as an auxiliary device for mathematical theory building, whereas engineering applications would provide at most heuristic inspirations for this. To be sure, von Neumann was well aware that mathematical modeling was feasible only within certain limits, but revealing these limits, he was convinced, would be a task for mathematics and logic itself. He admired, for instance, Gödel's incompleteness theorem as an outcome of superior logic delivering insights into the limits of mathematical reasoning.

4. The famous document about the EDVAC (von Neumann, [1945]1993) sketches what is nowadays called the von Neumann architecture.

5. See chaps. 16–18 of Ulam's (1986) book that are dedicated to von Neumann.

CONTROVERSY

Even during the planning stages of the Cybernetics Group, a predetermined breaking point was looming ahead, owing to the different views on the concepts of imitation, understanding, and modeling. Wiener projected cybernetics as a science that would investigate phenomena and models according to their functionality and behavior, not according to their material or inner structure. Ashby, oriented toward Wiener, stresses this point in his introduction to cybernetics:

> Cybernetics . . . does not ask "what *is* this thing?" but "*what does it do*?" Thus it is very interested in such a statement as "this variable is undergoing a simple harmonic oscillation," and is much less concerned with whether the variable is the position of a point on a wheel, or a potential in an electric circuit. It is thus essentially functional and behavioristic. (1957, p. 1; emphasis in original)

A function in this sense is multiply realizable, because the inner workings or the mechanisms that bring forth such behavior are not taken into account. In accordance with behaviorism, then fashionable in psychology, the issue was behavior without considering the mechanisms that produced it. Hence, Ashby uses a weak notion of functionalism. Wiener conceived of behavior and its control via feedback processes as one philosophical unit. In this respect, he treated human beings and machines in an analogous way. He backed this approach with a functionalist understanding of models. In a paper written jointly with Rosenblueth, Wiener emphasized what models should attain: "The intention and the result of a scientific inquiry is to obtain an understanding and a control of some part of the universe. This statement implies a dualistic attitude on the part of scientists" (Rosenblueth & Wiener, 1945, p. 316). Namely, there would be no guarantee that the attempts to gain control would also result in understanding, and vice versa.

In this context, these views lead to the distinction between "open-box" and "closed-box" (today more commonly called "black-box") approaches. The terminology stems from communication technology, in which closed-box procedures are employed to test a device by observing its input–output behavior without considering the inner electrical circuits or other mechanisms. In general, open-box models are in opposition to black-box models. The former represent in a more or less detailed way the structures and mechanisms of the phenomena being modeled, whereas the latter work merely functionally; that is, they model behavior, but do not include any account of the inner dynamics of the imitated phenomena. One can capture this difference by discerning *strong* and *weak* standpoints.

Whereas weak functionalism includes multiple realizability, it leaves open the question of mechanisms and their concrete implementation. All that matters is successful control. The strong standpoint, in contrast, is geared toward a deeper understanding that could hardly be achieved without specifying mechanisms. Though Wiener preferred the strong approach in principle, he also maintained that it holds little promise because of the limited epistemological capacities of human beings. Wiener distributed his preferences unequivocally: control takes priority over understanding whenever the complexity of an application context requires it.[6]

With regard to the aforementioned dualism, John von Neumann took a contradictory stance to Wiener and associated simulation with a completely different modeling strategy that should support the strong approach. This becomes clear in a long letter that von Neumann wrote to Wiener in 1946—that is, during the time they were both engaged in joint project planning.[7] In this letter, von Neumann criticized the choice of the human central nervous system as the primary object of study, saying it was too complex. He proposed changing the research agenda of the Cybernetics Group. Quite conforming to the classic strategy of reducing

6. For this reason, Galison (1994) is, after all, overstating the case when he brings together black-box modeling with an "ontology of the enemy"—an enemy who counts as inscrutable because he is malicious. Galison suggests a general connection between a warlike motivation and black-box modeling. In my opinion, the issue is more neutral—namely, an epistemological argument proposing that a weakly functionalist modeling can serve control (in the sense of accurate predictions) even when the model imitates only on the level of performance.

7. This letter was discovered in Wiener's estate relatively recently and was first published in Masani's (1990) biography. This is why it did not yet play a role in seminal older publications on von Neumann and especially in Aspray (1990).

complexity, von Neumann recommended taking viruses as objects better suited for the planned investigation. In the letter, he tried to convince Wiener that a virus is, finally, a cybernetic object in Wiener's sense, showing goal-oriented, feedback-controlled behavior. The letter, however, also documents a diverging conception of modeling.

> What seems worth emphasizing to me, however, is that after the great positive contribution of Turing–cum–Pitts–and–McCulloch is assimilated, the situation is worse than before rather than better. Indeed, these authors have demonstrated in absolute and hopeless generality, that anything and everything Brouwerian can be done by an appropriate mechanism and specifically by a neural mechanism—and that even one, definite mechanism can be "universal." Inverting the argument: Nothing that we may know or learn about the *functioning* of the organism can give, without "microscopic," cytological work, any clues regarding the further details of the neural mechanism. (Masani, 1990, p. 243; emphasis added)

Von Neumann alludes to Turing's results on computability and the universality of the Turing machine, which is a paper version of the digital computer, and also to artificial neural networks (ANNs). McCulloch and Pitts, who also participated in the Cybernetics Group, had shown that ANNs could, in principle, work as universally as a Turing machine. Finally, "Brouwerian" refers to the mathematician L. E. J. Brouwer (1881–1966), whose "intuitionism" required restricting mathematical procedures to the countably infinite. Of course, this holds for Turing machines, ANNs, and electronic computers in any case.

What is remarkable about this passage by von Neumann is that it basically contains a complete renunciation of Wiener's conception of cybernetics. It is precisely *because* simulation models are so adaptable that one cannot learn about the original mechanisms from a successful imitation of behavior ("functioning"). One should keep in mind, however, that although the adaptability of ANNs is secured theoretically, it has been more relevant in principle than in practice. Proving some ANN exists for any arbitrary behavioral pattern is a very different task from detailing one particular ANN that shows particular patterns of behavior. At least in principle, the considerations hit the nail on the head: the path leading from a modeling strategy oriented toward functions to a strategy oriented toward mechanisms might be practically impossible.

In chapter 3, I analyzed the plasticity of simulations in terms of structural underdetermination, a characteristic feature of simulation modeling. Right from the start, it blocks the inference from successful imitation to the validity of modeled mechanisms. Here it can be seen that combinations can turn out to be quite asymmetrical: a theoretically strong account can be tailored into a certain context of application by applying instrumental and pragmatic measures, whereas it seems hardly promising to start with a functional, instrumentalist approach and then try implanting a theoretical "backbone." Wiener, we might presume, would take this as supporting his point of view: imitating a complex phenomenon does not depend too strongly on simulating the mechanisms that generate it.

Von Neumann, on the other hand, took the very same consideration not as a confirmation but as a counterargument to Wiener—that is, as a *modus tollens* instead of a *modus ponens*. Without recognizing the principal conflict with Wiener, or at least without mentioning it, von Neumann suggested taking the possibility of structural modeling as the criterion and motivation for planning their research. In fact, he wanted to abandon the human nervous system as an object of study and replace it with a virus.

> Now the less-than-cellular organisms of the virus or bacteriophage type do possess the decisive traits of any living organism: They are self reproductive and they are able to orient themselves in an unorganized milieu, to move toward food, to appropriate it and to use it. Consequently, a "true" understanding of these organisms may be the first relevant step forward and possibly the greatest step that may at all be required.
>
> I would, however, put on "true" understanding the most stringent interpretation possible: That is, understanding the organism in the exacting sense in which one may want to understand a detailed drawing of a machine—i.e., finding out where every individual nut and bolt is located, etc. (quoted in Masani, 1990, p. 245)

The goal was a "real" and "true" understanding of the object or system under study, and this requires working with structurally isomorphic mechanisms. Put in other words, von Neumann adopted a mechanistic approach as the basic strategy for his research, and he was willing to modify the target of modeling in any way that made this strategy look promising. He focused on epistemic transparency, on ideal intelligibility attained via mathematical modeling—and the latter would be guided by this focus. The goal was an *open box* together with a blueprint; if simulation were to be helpful when doing this, it would be welcome.[8] In short, von Neumann clearly favored the strong viewpoint of modeling and regarded the capability of imitation and adaptation as a criterion for exclusion. For Wiener, on the other hand, exactly this capability was the reason for his soaring expectations for cybernetics.

Wiener did not answer this letter directly. The controversy over the conception of simulation modeling was something like a predetermined breaking point for von Neumann's and Wiener's joint project. Basically, both set their own agendas: Wiener elaborated the vision of cybernetics, whereas von Neumann became interested in (among other things) the numerical treatment of hydrodynamic equations, with meteorology as a paradigm case.

ATMOSPHERE: LAWS OR PATTERNS?

This case just described is particularly revealing because it shows how each of the two contested modeling strategies run into each other. From the start, von Neumann was convinced that meteorology, modeling the global circulation system, was ideally suited as a test case for new computational methods. It was based on theoretically mature mathematical (hydrodynamic) equations that nonetheless built a practically insolvable system. Von Neumann's initial hopes had been that simulations would deliver heuristic hints leading to a new theoretical mathematical approach to solving partial differential equations. This project was initiated by a

8. Von Neumann explicitly mentioned simulation experiments as potential aids.

working group at the IAS, headed by Jule Charney. The group made a first attempt to develop and apply computer code to meteorology and to work with the newly installed computing machine at the IAS. Needless to say, an exact numerical forecast of weather phenomena would be extremely difficult, if not impossible. However, von Neumann remained characteristically optimistic:

> Nonetheless, von Neumann's attitude was to push the pure-dynamics approach to the limit by carrying the computation through in the best possible way. . . . The decaying accuracy of the prediction, the inevitable decay with time, was fought or ignored insofar as possible. (Heims, 1980, pp. 152–153)

Charney's group upheld von Neumann's program: First, determine the relevant laws and mechanisms represented in a system of partial differential equations. Then, replace this system with a discrete system of difference equations (on a suitable grid). Finally, solve the latter system numerically with the computing power of the electronic machine. It is hardly surprising that Wiener was very critical of this program. He himself also viewed meteorological phenomena as paradigm cases, but in the sense of analyzing statistical patterns. For him, such analysis was not a makeshift but, rather, the only way to proceed.

Wiener was therefore quite unmistakably opposed to von Neumann's "solution" approach:

> It is bad technique to apply the sort of scientific method which belongs to the precise, smooth operations of astronomy or ballistics to a science in which the statistics of errors is wide and the precision of observations is small. In the semi-exact sciences, in which observations have this character, the technique must be more explicitly statistical and less dynamic than in astronomy. (Wiener, 1954, p. 252)

Wiener acknowledged that exact sciences—namely, astronomy and ballistics—benefit from a strong mechanistic-dynamic approach, whereas other sciences would be led astray when following the same line. The semi-exact sciences to which, according to Wiener, meteorology belonged, should follow a different approach right from the start. In his later autobiography, Wiener attacked the von Neumann and Charney group quite directly:

> Recently, under the influence of John von Neumann, there has been an attempt to solve the problem of predicting the weather by treating it as something like an astronomical-orbit problem of great complexity. The idea is to put all the initial data on a super computing machine and, by a use of the laws of motion and the equations of hydrodynamics, to grind out the weather for a considerable time in the future. . . . There are clear signs that the statistical element in meteorology cannot be minimized except at the peril of the entire investigation. (Wiener, 1956, p. 259)

This triggered a controversy in meteorology at the same time as the computer was being introduced as a new instrument: a controversy between statistical and dynamic approaches that persists to the present day. On first view, the solution approach à la von Neumann seems to have achieved a great success with climate science in exactly this contested area. However, the closer analysis in chapter 1 has revealed that the dynamic approach has been augmented in crucial points by instrumentalist components—and such components are oriented toward behavior rather than mechanisms. The long-term stability of simulated atmospheric phenomena was first attained in the 1960s through Arakawa's work, which dropped the ideal of solving the equations and, instead, added counterfactual assumptions justified on the level of performance (not mechanisms). The later success therefore hinges on a combination of von Neumannian and Wienerian or mechanistic-dynamic and functional approaches. This observation, I would claim, is not limited to the specific conditions of meteorology but, rather, is of general significance for simulation modeling.

CONSOLIDATION THROUGH COMBINATION

In their pure form, both modeling approaches are confined to a limited area of application. I would like to propose the following combination thesis: the success of simulations is based to a substantive degree on *combining* both conflicting approaches. Alternatively, one could speak of amalgamation, because this term indicates a robustness but a lack of homogeneity.

After all, one can speak of a numerical solution and refer to the strict sense of solution when, for instance, a zero of a complicated term is approximated by a computer algorithm with known error bounds.[9] Computation procedures designed for approximating by hand already present instructions for numerical solutions. In these cases, nothing except making work easier and faster seems to depend on electronic computers and simulation. It would be a fallacy, however, to derive a characterization of simulation from these cases, because in practice they are exceptions and not the rule.

Neither solution (in the strict sense) nor imitation can develop any great potential on its own. If one understands solution in the strict mathematical sense, one has to accommodate the fact that the mathematical instruments involved are incommensurable. The computer can execute only discrete, stepwise procedures, whereas traditional mathematized representations work in a continuum. However, a theoretical mathematical access to discrete dynamics has proved hard to attain; therefore, such dynamics have to be explored through simulation experiments.

In general, the iterative and exploratory mode of simulation modeling renders it difficult to keep track of how far the model depends on the original (mathematical) formulation, because the model is modified repeatedly on the basis of intermediate test evaluations. A completely different strategy is to reformulate the task in a way that drops out the extra approximation step—and hence the incommensurability. Simulation models then stand on their own and define the task from the start as a (mathematically) discrete one. Large parts of artificial life or artificial societies take this approach; that is, they investigate simulation models in their own right. However, completely independent of simulations, it is widely accepted that scientific agendas are influenced strongly by values,

9. Although this situation is, in a sense, favorable, the approximation character of numerical solutions still results in interesting complications (see Fillion & Corless, 2014).

instruments, and interests (see Carrier, 2010, among a host of others in philosophy and history of science who focus on this topic). The strategy of task transformation would be a case in which the instrument at hand (simulation) acquires a quasi-definitional character; that is, it decisively channels the research questions and projects.

On the other hand, strategies that concentrate exclusively on imitation often lack clues regarding what could be an efficient form of adaptation. As a result, practically all models invoke at least some structural assumptions. A well-known weakness of ANNs, for instance, is the notorious difficulty in finding adequate adjustments (effective learning algorithms).[10] Normally, the potential of ANNs can be operationalized only in models that work with a preset partial structure. This then serves as a starting point for parameter adjustments. Finally, it seems to me that by undervaluing structure, cybernetics has failed in its ambition to shape a new era in science. Both types of simulation approaches discussed here—von Neumann versus Wiener—present what could be called visions in pure forms or ideal types in the sense of Max Weber. They can be turned into successful applications only when combined (in varying proportions).

Meteorology is a paradigmatic example of how consolidation is achieved via combination. As explained, only a mixed form of von Neumannian (solution-oriented) and Wienerian (imitation-oriented) parts brought about success. Giving further examples is easy; however, the following short list might suffice.

1. Keller (2003) states that simulation consolidated in the 1960s. She bases this on the upswing of modeling with cellular automata (CA). I do not agree with assigning CA models such an elevated role when considering simulation methods. It is more the case that CA fit into the general picture. They also amalgamate the two types of simulation orientations and fit into my proposed combination picture. The way cells are arranged spatially and the way interactions between neighbors are specified are often

10. Recent accounts of data-driven research in "deep learning" claim to make inroads into the learning problem. This does not invalidate the combination thesis, as the discussion following will show.

guided by theoretical considerations about mechanisms. One instance is the (stochastic) dynamics of spiral galaxies discussed in chapter 2. There, the incommensurability gap became effective. Namely, theory presented the dynamics in (mathematically speaking) continuous terms that had to be expressed adequately in a discrete CA model. This required a modeling process of its own. Adaptable parameters, such as the speed of cell refinement, are employed to steer the modeling process. Such parameters cannot be determined by theory, nor by parameter-specific measurement. They can be determined only instrumentally by looking at the overall model outcome and then utilizing the available flexibility to find a sufficient adaptation. This process requires iterated simulation experiments. In the galaxy example, the visualized results had been crucial for judging how closely the model fitted the measured data—a step that did not involve analysis of mechanisms and therefore can be seen as a Wienerian component.

2. Recent claims regarding big data and the purported efficiency of deep-learning algorithms challenge the combination thesis. "Deep learning" refers to ANNs that include intermediate ("deep") layers. Owing to the vastly increased availability of learning data, together with high computational power, such models have made remarkable progress in pattern recognition and related problems such as face recognition and automated translation. The slogan is that sufficiently large databases will render theoretical accounts superfluous. Data-driven and big-data approaches give a name to an interesting trend in recent science, but heralding "the fourth paradigm" (Hey, Tansley, & Tolle, 2009) or "the end of theory" (Anderson, 2008) seems to be premature. First, data-driven approaches run into complications through spurious results and misclassifications that are hard to analyze. Calude and Longo (2017, p. 595) argue that most correlations in large databases are actually spurious; and as a result, the "scientific method can be enriched by computer mining in immense databases, but not replaced by it." The feasible way forward would thus be a combination. Second, analyzing the

methods of data analysis is itself a field that calls for mathematical modeling. Hence, data-driven science is a new challenge for theory-based modeling such as finding low-dimensional (computationally efficient) support for high- dimensional data (see Jost 2017).

3. What are termed "fingerprint methods" in climate science (Hegerl, et al., 1997) conceive of the temperature distribution around the globe as a pattern and evaluate it via statistical pattern analysis (much like common fingerprint detection). The issue is whether the observed patterns can be discriminated from simulated patterns emerging from models that do not take any human influence into account. If they can be discriminated, this is something like an "anthropogenic fingerprint." This ingenious approach grafts statistical pattern analysis (taking into account the complex dynamic interactions merely as the net effect as it appears in the temperature distribution) on a von Neumannian stock—that is, the theoretical model of atmospheric dynamics. Without this theoretical dynamic model, one could not (hypothetically) remove human influences and simulate how temperature would have developed without these influences. Statistical pattern analysis (which I subsume under the heading of simulation methods) proved key to arriving at a judgment on human influence.
4. In hurricane forecasting, researchers use general meteorological models and augment them with specially tailored parameterization schemata that gear them toward simulating hurricanes. In chapter 3, I discussed the case of Hurricane Opal and how black-box schemata proved to perform better than microphysical assumptions without any physical interpretation of parameters. Performance there meant simulating local behavior—that is, the path of the storm. The higher-level meteorological model is itself a combination of theoretical, pragmatic, and technological influences. Nonetheless, theoretical mechanisms (movement, thermodynamics) play an important role, but are combined with a functionalist black-box component.

5. Recent approaches try to find ways of coping with the flood of data produced by computer-based analyses of DNA. One such approach attempts to decode genome data via image-processing methods (Kauer & Blöcker, 2003). This time, Wienerian procedures are applied first to filter (supposedly) unimportant genetic noise. Then, in a second step, the reduced dataset is subjected to a fine-grained analysis oriented toward structure rather than statistics.
6. Artificial neural networks (ANN) have been introduced and discussed as a model class that can exhibit universal behavior. This means that such models can imitate phenomena without reproducing their dynamic mechanism (this fact motivated von Neumann's letter to Wiener). They "learn" adequate behavior with the help of a feedback loop that modifies internal weighing factors in such a way as to produce the desired performance. There are some cases in which the architecture of ANNs should represent certain target objects, but in the typical case, modeling via ANNs is motivated by the option of entirely neglecting any considerations on mechanisms in the target system. The work of Helge Ritter's robotics group at Bielefeld University illustrates this nicely. The group is combining a robot arm that corresponds very accurately to human physiology with a control unit modeled by an ANN. Their goal is to simulate the movements of a human arm when grasping objects such as a cup of coffee. The group has succeeded in building a surprisingly isomorphic robot replica (see figure 6.1), in which the ANN control unit utilizes the plasticity of the simulation model. It had to take this approach, because there is simply still no detailed knowledge on the effective mechanisms involved in how the human brain controls specific movements.
7. Social simulations, or agent-based models (ABMs), will serve as my last example in this somewhat contingent series. Developed from the pioneering work of Thomas Schelling (among others), they nowadays form a modeling school of their own (however, not confined to the social sciences, even reaching to philosophy and other areas). Quite generically, ABMs start by specifying

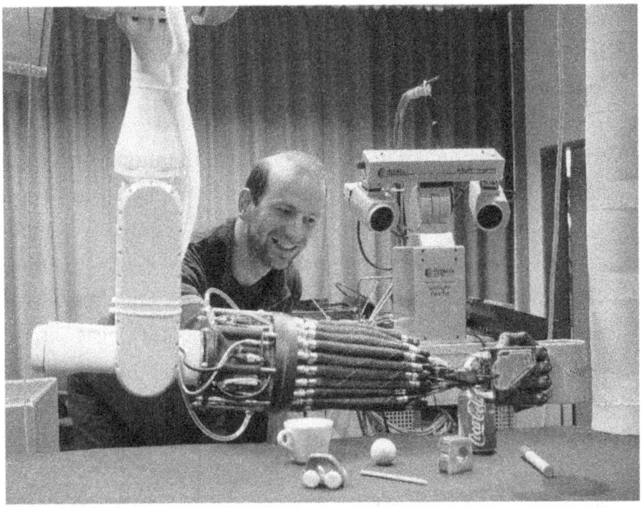

Figure 6.1 A robot arm grasping a cup of coffee in front of a researcher's arm. *Credit: WG Neuroinformatics of Bielefeld University. Courtesy of R. Haschke.*

plausible interactions among agents and then go through a phase of experimental adjustment and modification to navigate the models into an area in which they might have some significance. Schelling's famous model of social segregation encompassed only a few rules, but it took much patience and trial and error to calibrate the rules and parameters so that the model actually exhibited some interesting dynamics. In Schelling's case, the interesting phenomenon was segregation driven by a very moderate aversion to a superior number of "different" neighbors; that is, segregation took place in the model even on the assumption that everybody would prefer a mixed neighborhood. Evidently, such a model should employ rules that allow some plausible interpretation as social interaction. At the same time, there is an inevitably functionalist element at work, because the rules and parameters have to be modified and adjusted in light of the overall performance. Their mutual interaction is opaque, in the sense that it cannot be judged beforehand. This means that the decisive criterion is not whether some modeled representation of social rules is plausible but, rather, whether the rules interact in a way that produces phenomena that plausibly look like social phenomena.

8. The combinatorial nature is in no way restricted to model classes with high structural underdetermination. Thermodynamics is a field of physics that has a strong base in theoretical knowledge. There, equations of state provide exemplars of universally applicable general laws. The ideal gas law is the best-known example. This law makes it possible to determine important physical and chemical properties, but it works only in the "ideal" gas limit—that is, when molecules practically do not interact. There are also more elaborate equations of state, such as the van der Waals equation. However, when seeking an accurate prediction of chemical properties of substances in nonideal conditions (e.g., under considerable pressure as is typical for absorption columns used in industry), researchers have to work with equations of state that include several dozen adjustable parameters. Many of them do not have a physical interpretation but help to control the behavior of the equation. Adjusting these parameters in an explorative way is necessary when utilizing the theory-based equation (Hasse & Lenhard, 2017). Thus even in the most venerable field of thermodynamics, application-oriented simulations combine solution and imitation approaches.

This observation has far-reaching implications for simulation modeling. Any strategy for dynamic modeling that is oriented toward mechanisms has to demonstrate its performance on a *local* level. However, the systematic adjustment of local parameters and other parts of the model, which I discussed previously as following a "backwards" logic, is oriented toward the *global* level. This indicates that simulation modeling combines strong and weak approaches (in the specified sense). A characterizing feature of a particular simulation modeling approach would thus be what the combination looks like—that is, how the modeling process introduces instrumental components into the simulation and stabilizes them. In other words, the characterization would spell out what a simulation looks like in terms of the iterative and exploratory mode of modeling introduced in chapters 1 through 5.

This chapter has discussed two types of modeling—namely, a dynamic and a functional type that were opposed to each other in the pioneering days of simulation. As I have shown, both types can be combined in multiple ways. After the pioneering phase in the 1940s and 1950s, simulation modeling in a way consolidated and was seen as a feasible undertaking in different areas of science. The conflicting ideals of explanation via transparent model mechanisms versus universal applicability via highly plastic models were combined. Neither was maintained in its pure form. Simulation modeling as a new type of mathematical modeling involves the amalgamation of solution and imitation. How these two components are calibrated against each other varies widely across simulations.

I would like to point out that this combination thesis does not claim that a combination of strategies would present a panacea. First, combining antagonistic elements generally leads to a loss in consistency. Second, there is no guarantee of practical success. Even when growing theoretical understanding and computational power join forces, this does not automatically imply success. Success (also) depends decisively on the nature and complexity of the situation—fusion research might provide an example (see Carrier & Stöltzner, 2007). In general, complexity marks the limitations of the computer: simulation can sound out strategies regarding how to utilize the resources for building complex models while having to respect the boundaries of computational complexity.

IS EVERYTHING JUST IMITATION?

Even before computer simulation gained currency as an instrument in the sciences, philosophy of science was debating simulation in the context of artificial intelligence (AI). I would like to show that this older debate in AI exhibits the same tension between "strong" and "weak" approaches that I have just discussed. Regarding this point, the debate concerning AI in the philosophy of mind could profit from the philosophy of computer simulation.

One of the founding documents of AI is Alan Turing's (1950) essay "Computing Machinery and Intelligence," in which he introduced what later became known as the "Turing test" that he designed as an *imitation game*. One person types questions into a terminal and receives answers from two partners—namely, from another person and from a computer program. Can the first person decide, based on some length of conversation, which answers are from the human and which from the computer? A computer, or a computer program, whose answers would imitate those of a human being almost indistinguishably would pass the test and therefore qualify, per Turing's proposal, as being intelligent.

This test offers a sophisticated arrangement for filtering out functional equivalence—a computer should behave like a human, but the potential success should simultaneously be independent of the mechanisms implemented in the simulation model. Hence, Turing favored a "weak," functionalist strategy. The computer should simulate human behavior in the sense that Felix Krull (Thomas Mann's character in the unfinished novel *Confessions of Felix Krull, Confidence Man,* who feigned illness and other things) simulated illness—by matching the (known) relevant symptoms.

This standpoint has been criticized because it assumes weak criteria for defining and measuring success. Simulation indeed has some pejorative connotations of deceit. Contrary to this position, the "strong" approach of AI maintains that it is necessary for the model to represent the mechanisms and the structural organization of the target system before one is in a position to talk of intelligence. This controversy parallels the typology of simulation approaches investigated earlier and is known as the debate between strong and weak AI.

Keith Gunderson introduced the term simulation into the debate on AI in his talk "Philosophy and Computer Simulation" at the first PSA conference in 1968. However, he did not use the term in the weak (Turing–Krull) sense but rather, reserved "simulation" for the strong meaning: "the problem . . . of discerning when one subject (a machine) has *done the same thing* as another subject (a human being). And here 'doing the same thing' does not simply mean 'achieved similar end result'" (1985b, p. 96; emphasis in original).

Gunderson distinguishes "strong" cognitive simulation from "weak" AI along this criterion:

> As an exercise in AI, successful pattern-recognition (with respect to given characters of the alphabet) would consist in emulation of the results of exercising this ability. As an exercise in CS [cognitive simulation], a successful pattern-recognition program would simulate the process by which that ability was exercised. (1985b, p. 96)

Without doubt, Gunderson favors the strong position and understands simulation as the simulation of a dynamic process that is specified via mechanisms. A warning is necessary here, because the philosophy of mind does not use the term "simulation" in a consistent way. It invokes the notion in both its weak and its strong senses. In the quote just given, for instance, Gunderson had to shift to "emulation" to avoid ambiguity. Moreover, in the philosophy of mind, functionalism can denote a strong (in our sense) position. Jerry Fodor or Ned Block (Block & Fodor, 1972) take it as a position that refers to dynamic processes (instead of mere results).

More is at stake than a mere question of definition. Gunderson gives two arguments for why only the strong variant is adequate. Both arguments appear a couple of times with some slight variation in the literature; and they are particularly relevant, because they touch upon computer simulation in a quite general way. The first point, which is the heart of the matter for Gunderson, concerns robustness. The (weak) "net-result" approach would be wrong, according to Gunderson, because one can expect to attain good results under different conditions only by trusting the (adequately modeled) mechanisms that produced the results under certain conditions. This is, in fact, a strong argument against simulations that take insufficient account of theoretical and general assumptions.

I would like to add two points. First, Gunderson criticizes Turing on this account, but he has not fully taken into account how strong Turing's test actually is, because the imitation game potentially covers a wide range of verbal behavior. Any machine passing this test would have an impressive array of language competencies at its command. Although the Turing test radically reduces intelligence to language, speaking competently is a quite general ability. The second point I would like to add is about robustness. In most practical cases, the validity of a model or design is tested via rigorous technical procedures that address its performance; that is, it is done in a functionalist manner. Gunderson might perhaps doubt that any weak strategy could ever pass the test. This may or may not be correct; however, if it passes the test, this is a strong indicator. Simulation modeling can draw on a rich repertoire of adaptation and variation strategies. How much theoretical background is necessary to attain high adaptability remains an open question.

Gunderson's second argument criticizes functionalism directly: no success of a weak simulation would solve the problem that is at stake—namely, to investigate human intelligence. Along these lines, one could view John Searle's "Chinese room" (Searle 1980) as an objection to any weak version of AI.

I suspect the strong–weak opposition is not a productive one for AI. Functionalism in the philosophy of mind works with an analogy between the human brain and the computer. My analysis of computer simulation, however, has shown that simulation is fundamentally based on a combination of strong and weak components. It is plausible therefore that AI, exactly because it highlights the brain–computer analogy, could profit from an analysis of computer simulation.

However, this would require some instrumental aspects to also play a role in the discussion, as well as in actual modeling approaches. This is exactly what seems to be happening in the newer movement of robotics-oriented, or "behavioral," AI (see Brooks, 1991, 2002; Pfeifer & Scheier, 1999; Steels & Brooks, 1995). It departs from reducing intelligence to language but, rather, gears itself toward other types of human behavior such as orientation in space or grasping (as in the Bielefeld group mentioned earlier; see figure 6.1). These approaches do not aim to concentrate on the "net result" but instead capture the full process. At the same time, control units typically work with functionalist models. I take this as a strategy for capitalizing on a "strong" dynamic structure while simultaneously extending it with functionalist components. The robot arm of Ritter's group is such an example of a strategy oriented toward the mechanisms (modeling muscles, etc.) of grasping movements, but also relying on "weak" functional components. If behavioral AI is to succeed, the Turing test would lose its standing as an adequate criterion for (artificial) intelligence because it does not leave room for combinatorial approaches.

7

Validation, Holism, and the Limits of Analysis

One of the central issues in the epistemology of simulation is how to validate simulation-based knowledge. Just about any scientific procedure has to deal with this kind of question, and simulation is no exception. Which criteria should be used to judge whether a simulation is valid or not? Right from the start, it is clear that validation should not be understood as dealing with ultimate correctness; that is, validation surely does not determine whether a model has everything perfectly right. If this were to be the goal, the bar would be set so high that every practical procedure could not help but dislodge it. Then, validation would no longer be a problem of great interest. Instead, I assume that validation is concerned with evaluating models in certain respects and for certain purposes. Hence, one has to expect that the validation problem will not allow a homogeneous treatment—too manifold are the goals of simulations and the contexts in which they are deployed. This said, what are the particular or even characterizing features of simulation modeling when it comes to the problem of validation?

Put very briefly, simulations not only make validation harder in the sense of increasing the scale but also pose a particular new problem—namely, one of holism. The first part of the present chapter boils down to the thesis that simulations simply extend or amplify the validation problem. Simulations include steps that are not part of traditional mathematical modeling, especially when dealing with the expression of theoretical models as algorithmic implementations running on a material machine. These extra steps might have an exploratory and iterative character that enlarges the task of validation considerably. From the point of view taken in this first part of the chapter, simulations are indeed "more of the same"—more problematic modeling steps requiring a number of extra conditions—but they arguably do not pose a conceptually new type of validation problem.

The scale of modeling, however, creates a problem in an indirect way when simulation models appear to be as rich as reality itself. Under these conditions, it sometimes requires an effort to remind oneself that the contact with some part or aspect of reality cannot be presumed but, instead, is something that validation has to establish. Robert Rosen (1988) has pointed toward this kind of problem:

> The danger arises precisely from the fact that computation involves only simulation, which allows the establishment of no congruence between causal processes in material systems and inferential processes in the simulator. We therefore lack precisely those essential features of encoding and decoding which are required for such extrapolations. Thus, although formal simulators are of great practical and heuristic value, their theoretical significance is very sharply circumscribed, and they must be used with the greatest caution. (p. 536)

Simulations and computer models in general are making validation difficult, because comparing modeled processes with their potential partners in the causal world is becoming increasingly indirect. This comparison might even erroneously appear to be dispensable.

Simulations then present a trap for validation, or they suggest the fallacy of thinking that simulation is a particular case, although it is not. With more comprehensive modeling steps, both the effort involved and the difficulty in performing a validation increase rather than decrease. This perspective, however, needs to be supplemented.

The second part of this chapter deals with a particular problem that emerges when *complex* phenomena or systems are to be simulated by models that are also complex. The global dynamics of the simulation model are then generated from interacting modules in such a way that local adaptation can control and steer them but forestalls any possibility of grasping them analytically. The validation of mathematical models, however, is oriented toward analytical understanding, or maybe even defines what analytical understanding means. Hence, Bill Wimsatt's question, "What do we do when the complexity of the systems we are studying exceeds our powers of analysis?" (2007, p. 75) is highly relevant for simulation models. The limits of analysis become problematic when complex interactions together with a modular design govern how the model behaves. Simulations then gain a particular twist with dramatic consequences. Modularity is the very basis for handling complex systems, but erodes for reasons inherent to simulation modeling. In a way, simulations undermine their own working basis; and, as a consequence, the problem of holism emerges that reveals the limits of analysis.

PROOF AND VERIFICATION

The problems of validation escalate in a stepwise manner. Applying the computer in mathematics itself poses a somewhat simplified case from a modeling perspective, because the entities and processes to be modeled are already of a mathematical nature. One famous, early, and also controversial instance of working with a computational model is the proof of the four-color theorem by Appel and Haken (1977).[1]

Although it solved one of the most prominent and challenging problems in mathematics, the mathematical community received this proof with reservation. This was because at a decisive point, Appel and Haken had implemented a voluminous case-by-case differentiation. Only a computer could work through all the cases. As a result, the proof was completed only when the programmed computer algorithm reported that all cases had been checked. This did not seem satisfying, because

1. The question was how many colors are needed to color an arbitrary map of the world in such a way that (all) neighboring countries are colored differently. There are easy examples in which four colors are required, but it was unknown whether this number is sufficient in all cases. Euler ingeniously reformulated this question into a problem of two-dimensional topology (graph theory), but was still unable to solve it.

such a simple proposition as "for every map, four colors are sufficient for displaying neighboring countries in different colors" apparently called for an elegant and intelligible proof. The proof, however, distinguished so many cases that they could not be checked by hand. And even worse, the proof did not give any clue as to *why* all the cases came out positively. Put briefly, this proof was not based on the kind of analytical understanding that normally characterizes mathematical proofs, but it worked with an automated mechanism showing that the claimed proposition holds.

Initially the proof was questioned from two sides: there was not just the missing analytical understanding; there was also still doubt regarding how certain the proof actually was. If the proof were to have been accessible for analytical understanding, the single steps could have been followed conceptually, and doubt would not have been cast on the result with claims that it was subject to possible programming errors. Alas, in the present case, there was only the positive report by the software. Of course, the conception of the complicated case distinction was correct, but did the computer in fact carry out what it was supposed to? Did the actually implemented software check on all cases without making any mistakes?

Any mistake in programming or hardware would cast doubt on the proof. Thus we face a kind of prover's regress: the adequacy and correctness of the program must each be proven separately. This is a distinct step in the framework of validation that is called "verification" in the language of computer science. Some pieces of software, however, have a logico-algorithmic nature and are therefore not accessible to mathematical proof, at least not to proof based on understanding. "Why" a particular computer program implements a certain conceptional scheme without mistake cannot be understood. Instead, it must be checked by going through the implementation symbol by symbol.

In the case of the four-color theorem, the correctness of the program was evaluated piecewise. One can inspect smaller pieces of code, one can test whether the program achieves the correct result in those cases for which these results are known beforehand, and one can use different computers to make sure the particular hardware does not play a role. In this way, the community's doubt dissipated and Appel and Haken's proof became commonly accepted. However, the critique of missing analytical understanding remained.

The argumentation about epistemic opacity (see chapter 4) already pointed out that issues in simulation modeling and computer proof show important parallels. Tymoczko (1979) recognizes the main disturbing feature about computer proof as being its lack of *surveyability*. Computer usage not only questions the outstanding certainty of mathematical results but also deprives mathematics of a core element of its method: analytical transparency (surveyability).

Simulation modeling is becoming more comprehensive, or "thicker," because a simulation model comprises several layers: the mathematical layer, its conceptional scheme in terms of software, and the concrete implementation. A new problem becomes part of the evaluation of simulation—namely, the problem whether, on the level of executable commands, the machine receives precisely those commands that correspond to the conceptional scheme.[2] After all, the computer is, according to its idea, a logical machine, and there are strong movements in computer science to solve this question in a formal way. Such a solution would be a method for calculating whether a given program is actually correct. Interestingly, it is controversial whether such verification is at all possible.[3] This much is clear: to the extent that verification cannot proceed analytically, evaluating computer models will be in two parts: a conceptual part and some quasi-empirical part that is oriented toward the performance of the machine.

COMPLEXITY, VERIFICATION, AND VALIDATION

Up to this point, I have addressed computer-assisted proof in mathematics. Simulation models add more complexity to this picture. These

2. I put aside the additional question whether a computer always does execute the commands of some program with perfect reliability. MacKenzie (2001) offers a highly readable account of problems and cases relating to this issue.

3. Influential computer scientists such as Dijkstra hold that verification is possible in principle. DeMillo, Lipton, and Perlis (1977), however, have argued convincingly—against the background of the continuing failure of comprehensive attempts at mathematization in computer science—that in practice, full verification cannot be attained. From a philosophical perspective, Fetzer (1988) affirms that the gap between formal proof and concrete implementation cannot be overcome on a material device.

models are mediators to a special degree when they aim to simulate complex phenomena on the basis of a theoretical description. This was the issue addressed in the first five chapters of this book. The business of modeling is expanding immensely in terms of the number of modeling steps, because this is what the exploratory and iterative mode requires. Modeling is also enlarged in terms of scope; that is, phenomena are being simulated that used to be out of reach for mathematical modeling.

Atmospheric dynamics is just one example illustrating how the exploratory mode of simulation modeling simultaneously covers theoretical foundations and instrumental measures. Both enter the process of validation together.

The following brief episode from nanoscience exemplifies how far experiment, simulation, and theory are intertwined and depend on each other. The scientific instruments that arguably have contributed most to the rise of nanoscience are scanning-tunnel and atomic-force microscopes. Both visualize the forces measured by a very sharp tip (of molecular dimension) that is scanning a probe. Hence, these devices transform arrays of forces into visualizations. Science and technology studies have scrutinized this transformation in considerable detail.[4]

These microscopes, however, produce a great amount of very noisy data. Although the visualizations often appear clear and distinct, the data have to be smoothed and interpreted before they can be processed into an image. This is in itself no trivial task: modeling based on accepted background assumptions works like a filter between noisy raw data and cleaned data. One important element on which such models are based is called the Evans–Richie theory on the elasticity of single molecules.

A couple of years ago, a mathematically motivated study stumbled upon an error in this theory. This study did not question the theory; indeed, it had assumed it to be correct. Its aim was to optimize the filtering process (Evstigeev & Reimann, 2003). The theoretically promised solution was then tested in experimental measurements (with atomic-force microscopy), but could not be verified despite many attempts (Raible et al., 2004). The persistent failure was traced back eventually to an error in

4. See Mody (2006) for an impressive example.

the Evans–Richie theory, and this led to a significant adjustment (Raible et al., 2006). The point of this episode is to show that modeling can affect what counts as input data. Hence, it illustrates how simulation models and measurement might depend on each other. This observation leads to a problem of holism that is typical of complex simulations. This will be the main issue in the second part of this chapter.

Here I shall stick to arguing about interdependence. The processes of verification and validation are also mutually interdependent. Although both concepts should be distinguished in a conceptual sense, they are interrelated in practice. Indeed, any attempt to answer questions of verification completely separately from those of validation would seem to be a hopeless task. This observation supports the earlier finding that the mediating work of models increases with the complexity of simulations. Modeling is "thicker" in the sense of involving more models, more modeled aspects, and a richer dynamics. At the same time, modeling is becoming more indirect, because "thick" simulation models require a long process of iterated adaptations.

ARTIFACTS

When discussing complex modeling, I need to consider a second facet that arises out of multiplying and enlarging the steps of mediation. What happens in the simulation itself then acquires such a complex nature that it can become an object of investigation in its own right. Questions about validity, therefore, have to be separated into two categories: Do they deal with statements about phenomena in the real world (or target system)? Or are the external representational relations of secondary importance compared to relations between parts or entities within the model?[5]

Social simulations can serve as examples. They are about phenomena that occur in simulated societies. In this context, the question is whether the phenomena observed in the model (i.e., in the simulation) have anything informative to say about real societies. Another, related question is how much the simulated phenomena depend on particular choices of

5. This differentiation is useful analytically even if there are concrete examples in which the difference is far from clearcut.

model parameters or if they are robust across a larger class of models. This latter question is about validation and asks whether a certain behavior or phenomenon typically appears in an artificial society—instead of being an artifact of the implementation.

On first view, this might appear a bit paradoxical, because these societies are in any case artificial *in toto*. However, what the term "artificial" means depends strongly on the context in which it is used. If, for instance, one is simulating fluid flow around a fixed obstacle, then the entire simulation is artificial in a very obvious sense—it is not the real flow. Now, consider that a turbulence pattern of a specific form emerges behind the obstacle. This could be a phenomenon due to the modeled regularities of fluid flow, or it could be due to particularities of the implementation such as the choice of grid. In both cases, the turbulence pattern is artificial, but it is artificial in importantly different senses.

Consequently, part of validating the model would be to check whether the patterns observed in the simulation are invariant under different discretizations. This question can be answered by performing an internal investigation of the model world. If the patterns cannot be reproduced by empirical experiment, they might be artifacts of the theoretical model assumptions. This would be a question external to the model. In general, one might express the task of validation as the task of detecting artifacts.

However, this approach is limited because, under certain conditions, simulations explicitly *require* artificial assumptions to create realistic patterns. If validation is equated with detecting and excluding artifacts, one needs the right and complete theory, as well as a model that implements this theory without any flaws. Any deviation would result in an artifact (in the rigorous sense I am discussing right now). Such a stance would take simulations as numerical approximation—something that I discarded as being too simplistic in chapter 5.

Furthermore, in fields such as *artificial societies* or *artificial life*, comparison with external phenomena does not play a major role. At the same time, however, there is hardly ever a theoretical model that is accepted as the right model. Consequently, validation does not deal with comparing the simulated dynamics with the dynamics of some (original) theoretical model. How much the "artificial" mechanisms represent relationships that hold in "real" society is a question of little concern in these fields. The rule is, rather: do whatever generates interesting phenomena. Nevertheless, questions of validation are also discussed in these fields.

Validation is then interpreted consistently as an "internal" affair of artificial societies that concerns how robust the description of artificial societies is—for instance, whether some phenomenon observed in a simulation follows on from how individuals interact (in the model) and is invariant under a variation of assumptions, or whether this phenomenon is an "artifact" of a very particular choice of parameter values and stops occurring when these parameters are varied.

One telling example of this kind of validation problem can be found in Thomas Schelling's work on social segregation mentioned in chapter 2. He designed a tableau of neighboring (unoccupied or occupied) cells, and each cell could assume one of two states, say x and y. The states could be interpreted as denoting whether a family of type x or y lives in this location or cell. The dynamics initially envisaged families moving to a previously unoccupied cell if the majority of neighboring cells are occupied by individuals (families) of the other type. Schelling started out with experiments on paper, and it appeared to him that social segregation (i.e., neighborhoods of merely one type) would emerge even under these rudimentary assumptions. But how valid is such a result (in the model world)? Answering this question required a large number of experiments that varied such conditions as the starting configuration of types in cells, the number of cells, and so on. To conduct these experimental explorations, Schelling needed a computer that automatized them. Although the dynamics of cells allows a simple formal description, it does not have a (closed-form) analytical solution. Handling a network of cells requires a computer. Hence, model dynamics of this kind are adapted to, or channeled toward, the computer.

This kind of *growing explanations* follows a bottom-up strategy. In other words, the aim is to simulate, or "grow," complex phenomena based on relatively simple assumptions such as those in social segregation. If patterns emerge that can be interpreted as social segregation, and if the model is indeed simple, then the model behavior provokes a reaction such as, "What—is it that simple?" Under these conditions, given that the test of robustness proves to be positive, validating the modeling assumptions and explaining the phenomenon happen at the same time. Norton Wise

(2004) identifies a new paradigm of explanation here. Validation and explanation are closely intertwined. Nonetheless, one should impose the constraint that explanation here means proposing simple assumptions whose success renders more complicated candidates obsolete. In a sense, this sort of explanation applies Occam's razor to the description of social phenomena.

This observation is not limited to social simulations but also holds for all cases in which the simulation model becomes the object of investigation in its own right. The complexity of this object (i.e., the simulation model itself) requires validation steps. In some ways, this is very similar to experimental results in the ordinary sense of the experiment. In the context of simulation, therefore, validation comes in various shades. Most of all, when speaking about validation, however, one has to specify how far one is dealing with relationships that are internal or external to the model.

GRADATIONS

Simulation models can be found in very different grades of complexity, and the problems of validation change with these grades. Simulations are not complex per se. There are very simple models in which specification and implementation can be of paramount transparency—think of Schelling's neighborhood dynamics or some percolation models in physics. The Ising model is maybe the most prominent among these. Locations (grid points) on a grid take on one of two possible states—for instance, spin up or spin down—and these states are influenced by the states at neighboring locations. With rising "temperature," the influence from neighboring states decreases and is replaced by an independent random choice. The Ising model is famous because it is a simple model that can, nevertheless, describe phase transitions. There is a critical temperature below which spontaneous clustering occurs when a cluster is a set of neighboring grid points that have the same state. Above the critical temperature, states become more independent and clusters dissolve. The fact that a critical temperature exists is a mathematical result that can be derived straightforwardly.

Determining the value of this temperature, however, is notoriously difficult. This is because a multitude of interactions between grid points are relevant at the same time (Hughes, 1999, provides an insightful description of this case).

Determining the critical temperature by simulations is much easier. One can run the model dynamics with different temperatures and simply "observe" which value is critical—that is, at (or around) which temperature clusters start to form. The Markov chain Monte Carlo (MCMC) methods can simulate the probability distributions that result from an impressive range of complex models, and they can do so in very moderate computation time.[6] Simulation experiments never deliver closed-form solutions, but their results can document how certain parameter choices affect the model dynamics. In cases in which the model behavior is derived independently from theoretical reasoning, as with phase transitions in the Ising model, the simulation models are validated (confirmed) if they actually generate this behavior.

Depending on the degree of complexity, solving (rather than simulating) the Ising model is extremely difficult. The problems of validation, however, are much simpler, because the model is of a discrete, computer-ready nature in any case, and the simulation dynamics are quite simple—apart from the many iterations that are necessary before the model reaches some stable state. However, these do not matter when the computer does them. Hence, this is the simplest grade for validation.

The case becomes more complicated when the simulation model is based on a theoretical model that is not of a discrete nature and hence cannot be implemented straightforwardly on a computer. Discretization might call for considerable conceptual effort, particularly as questions regarding what some adequate discrete formulation of the theoretical model looks like are typically "artificial" questions whose answers depend more on the instrument used and on the purpose at hand than on the theoretical model.

6. For an ingenious appreciation of how MCMC simulations affect statistical methods, see Diaconis (2009).

Often, simulation is the method of choice when a theoretical model is lacking or when various local models have to be brought together. Major car manufacturers forgo crash-testing real car bodies. They have replaced these tests with simulations that employ finite element models comprising all relevant structural parts of the (virtual) car body. These models have parameters that have to be adapted in a phenomenological way; that is, their values are assigned so that the simulated body reproduces the behavior of the material body to be found in those cases that are already known. If this condition is satisfied, the models count as adequate for new designs.[7] A similar case is models of evacuation in which the behavior of people leaving a building or aircraft can be simulated.[8] Such simulation models are not entirely void of theory, but can be validated only when compared to data on real-world behavior. This is how one would expect things to be, owing to the models' plasticity (chapter 3).

Because the model behavior is influenced decisively by the particular specification, the validation process has to operate with a continuity assumption—namely, that the choice of parameters does not depend strongly on the concrete case. Otherwise, the model would be worthless when it comes to other cases. This continuity assumption looks plausible only, I think, when there is a sufficiently general theoretical reasoning behind the model.

Then moderate extrapolations come within reach; that is, it becomes possible to transfer the model to cases in which there can be no comparison with empirical data. "Moderate" means these cases will have to be sufficiently close to known and tested cases for researchers to be able to trust that the simulation results will be reliable.[9] In these cases, the following two conditions arguably hold: First, confidence in certain approaches to simulation modeling grows over successful applications and empirical testing. Second, extrapolation stays in the vicinity of the known cases.

7. This is valid for both the design process and the construction process. Nonetheless, independent of how reliable the producers think their product is, official certification procedures normally test the material objects.

8. Dirk Helbing (1995) has done pioneering work in this direction.

9. The United States, for instance, agreed to ban atomic bomb testing after becoming convinced that their simulation models were of sufficient validity to produce and maintain nuclear weapons.

Model validity is confirmed in test cases and is then assumed to cover cases of a similar nature. Admittedly, the exact meaning of "similar" is hard to tell here and surely depends on, among other things, disciplinary conventions. For example, simulation software is employed in design processes for architecture and structural engineering. High-rise buildings are designed with the help of such software as long as the type of building is sufficiently close to types in which the design has been successful.

There are good grounds for objecting that this criterion of validity appears to be circular or at least vague. How can you explicate under which conditions something is of the same type without referring to the validity of some model assumptions? How can you clarify the continuity assumption? Answering these questions requires some theoretical foundation regarding when a model is valid. I think this is an open problem that is relevant for scientific practice and can be approached only in an empirically based, stepwise, and exploratory manner. Failures are therefore included—simulations cannot be turned into an infallible instrument. Quite the contrary, procedures of model validation are riddled with mistakes and errors. A prominent (because officially investigated) example happened in 1991 when the drilling rig *Sleipner A* sank off the coast of Norway. The investigation concluded that a mistake in the simulation software was responsible for the accident. More precisely, the structural engineering calculations worked with a discretization procedure that was inadequate at certain neuralgic points, and this resulted in an overestimation of the rig's stability. There is a well-known discussion in engineering design about how fallible criteria of validation actually are in practice (see Petroski, 1992, for a general account and Peterson, 1995, for a particular focus on software errors).

The series of gradations now arrives at cases such as Landman's nanowire (see chapter 2). The surprising phenomenon in the simulation was later confirmed by a specially designed empirical experiment. Such success not only shows that Landman used an adequate simulation model but it also highlights that experimental confirmation was seen as necessary to validate the simulation result. Landman himself emphasized the need for close ties between simulation modeling and laboratory experiments. At this point, validation starts to become controversial, because simulation models reach out to the (yet) unknown. Küppers, Lenhard, and Shinn

(2006) argue that comparing simulations with empirical data remains indispensable for validation. The critical cases are those in which a comparison to real-world phenomena is no longer possible. These cases, however, are the particularly interesting ones for simulation modeling, because there might well be no alternative.

Eric Winsberg (2006a) examines exactly these cases. Simulations, he concludes, can achieve "reliability without truth." One has to dispense with truth, because what he calls the "fictional" (I would prefer to speak of "instrumental") components do not allow an isomorphic relation between model and reality. On what grounds, then, can models be called reliable?

Winsberg refers to modeling techniques that transfer the validity they have acquired in previous cases. This once again raises the question mentioned earlier: In which sense are other (previous) cases relevant here? If one cannot specify some plausible reasons for when cases are and when they are not of a similar nature, validity would seem to be little more than a vague hope or audacious claim.

This diagnosis holds particularly when highly adapted models are involved that exhibit little unificatory power. If simulation models have been adapted on the basis of their high plasticity, then referring to previous successes of the modeling techniques does not contribute much to confirming validity. Put differently, simulations cannot derive their validity from the success of the techniques they employ, because the behavior of these simulation models does not depend on the modeling technique alone. It also depends on the methods of adaptation that are likewise a systematic part of simulation methodology (which restricts the scope of Winsberg's claim). Simulation-based knowledge therefore appears to be questionable when there is neither a theoretical background with some unificatory power nor rigorous testing against real-world counterparts of the simulated phenomena.

In short, there is no simulation-specific bootstrap available that could pull up and elevate simulations to high validity when empirical tests are lacking. Earlier, I argued that iterativity is a strategic means of solving modeling problems that had been formerly out of reach. These problems, however, are of a *methodological* nature and leave the question of validity largely untouched. I do not see any particular reason why simulation techniques validated in one field can be transferred to another field and remain valid there. Of course, they *can* remain valid, but this would require some additional argument. If some reliable theory serves as the basis of the simulation model and covers both fields, it could be the core of such an argument. In typical cases, however, simulations contain instrumental components that weaken the relationship to theory, so that the latter cannot grant validity. For this reason, so I think, the comparison with target phenomena remains the pivotal criterion. Ultimately, the more instrumental and the more exploratory the model is, the closer it has to match test cases in reality in order to validate its results.

HOLISM

The first part of this chapter boiled down to the observation that enlarging the realm of modeling also enlarges the problems with validation. Because simulation modeling often works with a particularly rich set of modeling levels, an obvious question is whether the (likely) increase in validation problems might at some time reach a critical point. In the second part of this chapter, which begins right here, I want to show that this is, in fact, what happens. Complex computer and simulation models are plagued with a problem of confirmational holism that highlights the limits of analysis. Through this holism, an analytical understanding of the model's dynamics becomes practically out of reach. Such an understanding would be indicated by the ability to attribute particular dynamic features of the simulation to particular modeling assumptions.

Imagine a scene at an auto racetrack. The air smells of gasoline. The driver of the F1 racing car has just steered into his box and is peeling himself out of the narrow cockpit. He takes off his helmet, shakes his sweaty hair, and then his eyes make contact with the technical director expressing a mixture of anger, despair, and helplessness. The engine had not worked as it should, and for a known reason: the software. However, the team had not been successful in attributing the miserable performance to a particular parameter setting. The machine and the software interacted in unforeseen and intricate ways. This explains the exchange of glances between driver and technical director. The software's internal interactions and interfaces proved to be so complicated that the team had not been able to localize an error or a bug. Instead, they remained suspicious that some complex interaction between seemingly innocent assumptions or parameter settings was leading to the insufficient performance.

This story happened in fact, and it is remarkable because it displays how invasive computational modeling is, and how far it has encroached on what would seem to be the most unlikely areas. I reported this short piece for another reason, however, because the situation is typical for complex computational and simulation models. Validation procedures, while counting on modularity, run up against a problem of holism.

Both concepts, modularity and holism, are notions at the fringe of philosophical terminology. *Modularity* is used in many guises and is not a particularly philosophical notion. It features prominently in the context of complex design, planning, and building, from architecture to software. Modularity stands for first breaking down complicated tasks into small and well-defined subtasks and then reassembling the original global task following a well-defined series of steps. It can be argued that modularity is a main pillar on which various rational treatments of complexity rest, from architecture to software engineering.

Holism is a philosophical term to some degree, and it is covered in recent compendia. The *Stanford Encyclopedia of Philosophy*, for instance, includes (sub)entries on methodological, metaphysical, relational, or meaning of *holism*. Holism generically states that the whole is greater than the sum of its parts, meaning that the parts are in such intimate interconnection that they cannot exist independently of the whole or cannot be understood without reference to the whole. Especially, W. V. O. Quine has made the concept popular, not only in the philosophy of language but also in the philosophy of science, where one speaks of the Duhem–Quine thesis. This thesis is based on the insight that one cannot test a single hypothesis in isolation but, rather, that any such test depends on "auxiliary" theories or hypotheses—for example, how the measurement instruments work. Thus, any test addresses a whole ensemble of theories and hypotheses.

Lenhard and Winsberg (2010) have discussed the problem of confirmation holism in the context of validating complex climate models by conducting model intercomparison projects. They argue that "due to interactivity, modularity does not break down a complex system into separately manageable pieces" (p. 256). In a sense, I want to pick up on this work, but place the thesis in a much more general context; that is, I want to point to a dilemma that is built on the tension between modularity and holism and that occurs quite generally in simulation modeling.

My main claim in this second part of chapter 7 is as follows: According to the rational picture of design, modularity is a key concept in building and evaluating complex models. In simulation modeling, however, modularity tends to erode for reasons inherent in simulation methodology. Moreover, the very condition for successful simulation undermines the most basic pillar of rational design. Thus the resulting problem for validating models is one of (confirmation) holism.

This second part of the chapter starts by discussing modularity and its central role for what is termed the "rational picture of design." Herbert Simon's highly influential parable of the watchmakers will feature prominently. It paradigmatically captures complex systems as a sort of large clockwork mechanism. I want to emphasize the disanalogy to how simulation modeling works. Simulation is based on an iterative and exploratory mode of modeling that makes modularity erode.

I shall present two arguments supporting the erosion claim, one from parameterization and tuning, and the other from kludging, or kluging. Both are, in practice, part and parcel of simulation modeling and both make modularity erode. The chapter will conclude by drawing lessons on the limits of validation. Most accounts of validation require modularity (if often not explicitly), and are incompatible with holism. In contrast, the exploratory and iterative mode of modeling restricts validation, at least to a certain extent, to testing (global) predictions. This observation shakes the rational (clockwork) picture of design and of the computer.

THE RATIONAL PICTURE

The design of complex systems has a long tradition in architecture and engineering. Nonetheless, it has not been covered much in the literature, because design has been conceived as a matter for experienced craftsmanship rather than analytical investigation. The work of Pahl and Beitz

(1984/1996/2007) gives a relatively recent account of design in engineering. A second, related source for reasoning about design involves complex computer systems. Here, one can find more explicit accounts, because the computer led to complex systems much faster than any tradition of craftsmanship could grow, thus rendering the design problem more visible. A widely read example is Herbert Simon's *Sciences of the Artificial* (1969). Right up to the present day, techniques of high-level languages, object-oriented programming, and so forth are producing rapid changes in the practice of design.

One original contributor to this discussion is Frederic Brooks, a software and computer expert (and former manager at IBM), as well as a hobby architect. In his 2010 *The Design of Design*, he describes the widely significant rational model of design that is (informally) adopted in practice much more often than it is explicitly formulated in theoretical literature. The rational picture starts by assuming an overview of all options at hand. According to Simon (1969, p. 54), for instance, the theory of design is the general theory of searching through large combinatorial spaces. The rational model, then, presupposes a utility function and a design tree that span possible designs.

Brooks rightly points out that these are normally not at hand. Nevertheless, design is conceived as a systematic, step-by-step process. Pahl and Beitz try to detail these steps in their rational order. Simon also presupposes the rational model, arguably motivated by making design feasible for artificial intelligence (see Brooks, 2010, p. 16). Wynston Royce (1970), to give another example, introduced the "waterfall model" for software design. Royce was writing about managing the development of large software systems, and his waterfall model consisted in following a hierarchy ("downward"), admitting the iteration of steps on one layer but no interaction with much earlier ("upward") phases of the design process. Although Royce actually criticized the waterfall model, it was cited positively as a paradigm of software development (see Brooks, 2010, on this point).

The hierarchical order is a key element in the rational picture of design and it presumes modularity. Let me illustrate this point. Consider first a simple brick wall (figure 7.1). It consists of a multitude of modules, each with a certain form and static properties. These are combined into

Figure 7.1 A brick wall as example of modularity. *Credit: own photograph.*

potentially very large structures. It is a strikingly simple example, because all modules (bricks) are similar.

A more complicated, though closely related, example is the one depicted in figure 7.2, in which an auxiliary building at Bielefeld University is being assembled by modular construction.

These examples illustrate how deeply ingrained modularity is in our way of building (larger) objects. A third and last example (figure 7.3) displays the standard picture in software design that is used generically whenever a complex system is developed.

Some complex overall task is split up into modules that can be tackled independently and by different teams. The hierarchical structure should ensure that the modules can be integrated to make up the original complex system. Modularity does not just play a key role when designing and building complex systems; it is also of crucial importance when taking account of the system. Validation is usually conceived in the very same modular structure: independently validated modules are put together in a controlled way to make up a validated bigger system. The standard account of how computational models are verified and validated gives rigorous guidelines that are all based on the systematic realization of modularity (Oberkampf & Roy, 2010). In short, modularity is the key for both designing and validating complex systems.

Figure 7.2 A part of Bielefeld University being built from container modules. *Credit: Courtesy of N. Langohr.*

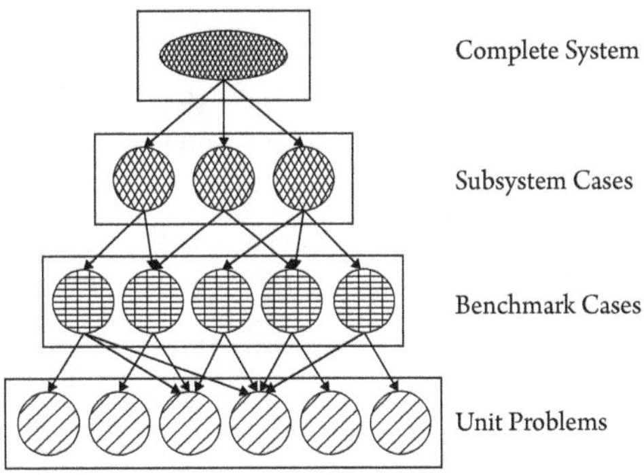

Figure 7.3 Generic architecture of complex software. *Credit: From AIAA Guide for the Verification and Validation of Computational Fluid Dynamics Simulations (1998), fig. 4, p. 11. Courtesy of American Institute of Aeronautics and Astronautics.*

This observation is expressed paradigmatically in Simon's parable of the two watchmakers. It can be found in Simon's 1962 paper "The Architecture of Complexity," which became a chapter in his immensely influential book, *The Sciences of the Artificial* (1969). There, Simon investigates the structure of complex systems. The stable structures, according to Simon, are the hierarchical ones. He expressed this idea by recounting the parable of the two watchmakers named Hora and Tempus (pp. 90–92). Philip Agre describes the setting in the following words:

> According to this story, both watchmakers were equally skilled, but only one of them, Hora, prospered. The difference between them lay in the design of their watches. Each design involved 1,000 elementary components, but the similarity ended there. Tempus' watches were not hierarchical; they were assembled one component at a time. Hora's watches, by contrast, were organized into hierarchical subassemblies whose "span" was ten. He would combine ten elementary components into small subassemblies, and then he would combine ten subassemblies into larger subassemblies, and these in turn could be combined to make a complete watch. (2003, p. 416)

Because Hora takes additional steps and builds modules, Tempus's watches need less time for assembly. However, it was Tempus's business that did not thrive, because of an additional condition not yet mentioned—namely, some kind of noise. From time to time, the telephone would ring and whenever one of the watchmakers answered the call, all the cogwheels and little screws fell apart and that watchmaker had to recommence the assembly. Whereas Tempus had to start from scratch, Hora could keep all the finished modules and work from there. In the presence of noise, so the lesson goes, the modular strategy is superior by far. In summarizing that modularity, Agre (2003) speaks of the functional role of components that evolved as a necessary element when designing complex systems:

> For working engineers, hierarchy is not mainly a guarantee that subassemblies will remain intact when the phone rings. Rather, hierarchy simplifies the process of design cognitively by allowing the functional role of subassemblies to be articulated in a meaningful way in terms of their contribution to the function of the whole. Hierarchy allows subassemblies to be modified somewhat independently of one another, and it enables them to be assembled into new and potentially unexpected configurations when the need arises. A system whose overall functioning cannot be predicted

> from the functionality of its components is not generally considered to be well-engineered. (p. 418)

Now, the story works with rather particular examples insofar as watches exemplify complicated mechanical devices. The universe as a giant clockwork has been a common metaphor since the seventeenth century. Presumably, Simon was aware that the clockwork picture is limited, and he even mentioned that complicated interactions could lead to a sort of pragmatic holism. Nonetheless, the hierarchical order is established by the interaction of self-contained modules.

There is an obvious limit to the watchmaker parable: systems have to remain manageable by human beings (watchmakers). There are many systems of practical interest that are too complex for this, from the earth's climate to the aerodynamics of an airfoil.

Computer models open up a new path here, because simulation models might contain a wealth of algorithmic steps far beyond what can be conceived in a clockwork parable.[10] From this point of view, the computer appears as a kind of amplifier that helps to revitalize the rational picture. Do we have to look at simulation models as a sort of gigantic clockwork? In the following, I shall argue that this viewpoint is seriously misleading. Simulation models differ from watches in important ways, and I want to focus on this disanalogy. Finally, investigating simulation models will challenge our picture of rationality.

EROSION OF MODULARITY 1: PARAMETERIZATION AND TUNING

In stark contrast to the cogwheel image of the computer, the methodology of simulation modeling tends to erode modularity in systematic ways. I want to discuss two separate though related aspects: first, parameterization and tuning; and second, kluging (also called kludging). Both are, for

10. Charles Babbage had designed his famous "analytical engine" as a *mechanistic* computer. Tellingly, it encountered serious problems exactly because of the mechanical limitations of its construction.

different reasons, part and parcel of simulation modeling, and both make the modularity of models erode. I shall investigate them in turn and develop two arguments for erosion.

Parameterization and tuning are key elements of simulation modeling that stretch the realm of tractable subject matter far beyond what is covered by theory. Or, more precisely, simulation models can make predictions, even in fields that are covered by well-accepted theory, *only* with the help of parameterization and tuning. In this sense, the latter are success conditions for simulations.

Before I start discussing an example, let me add a few words about terminology. There are different expressions that specify what is done with parameters. The four most common ones are: calibration, tuning, adaptation, and adjustment. These notions describe very similar activities, but also differ in how they define what parameters are good for. *Calibration* is commonly used in the context of preparing an instrument—for example, a one-off calibration of a scale to be used very often in a reliable way. *Tuning* has a more pejorative tone—for example, achieving a fit with artificial measures or a fit to a particular case. *Adaptation* and *adjustment* have more neutral meanings.

Atmospheric circulation, my companion since the first chapter, can serve again as an example. It is modeled on the basis of accepted theory (fluid dynamics, thermodynamics, motion) on a grand scale. Climate scientists call this the "dynamic core" of their models, and there is more or less consensus on this part. Although the theory employed is part of physics, climate scientists mean a different part of their models when they speak of "the physics." This includes all the processes that are not completely specified as from the dynamic core. These processes include convection schemes, cloud dynamics, and many more. The "physics" is where different models differ, and the physics is what modeling centers regard as their achievements and what they try to maintain even as their models change into the next generation.

The physics acts like a specifying supplement to the grand-scale dynamics. It is based on modeling assumptions about which subprocesses are important in convection, what should be resolved in the model, and what should be treated via a parameterization scheme. Often, such processes are not known in full detail, and some aspects depend (at least) on what happens on a subgrid scale. The dynamics of clouds, for instance, depend on a staggering span of very small (molecular) scales and much larger scales spanning many kilometers. Hence, even if the laws that guide these processes were known, they could not be treated explicitly in the simulation model for reasons of complexity. Modeling the physics has to bring in parameterization schemes.[11]

For example, how do cloud processes work? Rather than trying to investigate and model all microphysical details, such as how exactly water vapor is entrained into air (building clouds), scientists use a parameter, or a scheme of parameters, that controls moisture uptake in a way that (roughly) meets known observations. Often, such parameters do not have a direct physical interpretation; nor do they need one when, for example, a parameter stands for a mixture of (partly unknown) processes not resolved in the model. One then speaks of using an "effective" parameter. The important property of working parameterizations is not accuracy in representation on the small scale.[12] Instead, the parameterization scheme has to be flexible so that the parameters of such a scheme can be changed in a way that makes the overall model match some known data or reference points. Representation then becomes relevant on a much larger scale.

This rather straightforward observation leads to an important fact: a parameterization, including assignments of parameter values, makes sense only in the context of the larger model. Observational data are not compared to the parameterization in isolation. The Fourth Assessment Report of the Intergovernmental Panel on Climate Change (IPCC) acknowledges the point that "parameterizations have to be understood in the context of their host models" (Solomon et al., 2007, sec. 8.2.1.3)

The question whether the parameter value that controls moisture uptake (in my oversimplified example) is adequate can be answered only by examining how the entire parameterization behaves and, moreover, how the parameterization behaves within the larger simulation model. Answering such questions would require, for instance, looking at more

11. Parameterization schemes and their more or less autonomous status are discussed by Gramelsberger and Feichter (2011), Parker (2014b), or Smith (2002).

12. Of course, physically well-motivated, "realistic" parameterizations increase the credibility of a model. But if they are not available, more pragmatic parameterizations offer a way out.

global properties, such as mean cloud cover in tropical regions or the amount of rain in some area. Briefly stated, parameterization is a key component of climate modeling, and adjusting parameters is part and parcel of parameterization.[13]

It is important to note that tuning one parameter takes the values of other parameters as given—be they parameters from the same scheme or parts of other schemes that are part of the model. A particular parameter value (such as controlling moisture uptake) is judged according to the results it yields for the overall behavior (such as the global energy balance). In other words, tuning is a local activity that is oriented toward global behavior. Researchers might try to optimize parameter values simultaneously, but for reasons of computational complexity, this is possible only with a rather small subset of all parameters. In climate modeling, skill and experience remain important for tuning.

Furthermore, tuning parameters is not just to orient toward the global model performance; it also renders the local behavior, in a sense, invisible. This is because every model will be importantly imperfect, as it contains technical errors, works with insufficient knowledge, and so forth—which is just the normal case in scientific practice. Now, tuning a parameter according to the overall behavior of the model then means that the errors, gaps, and bugs compensate each other (if in an opaque way). Mauritsen et al. (2012) have pointed this out in their pioneering article on tuning in climate modeling.

In climate models, cloud parameterizations play an important role because they influence key statistics of the climate while, at the same time, covering major (remaining) uncertainties about how an adequate model should look (see, e.g., Stevens & Bony, 2013). Building parameterization schemes that work with "physical" parameters (which have a straightforward physical interpretation) is still a major challenge (Hourdin et al., 2013). In the process of adjusting the (physical and nonphysical) parameters, these schemes inevitably become convoluted. The simulation is then based on the balance of these parameters in the context of

13. The studies of what is termed "perturbed physics" ensembles show convincingly that crucial properties of the simulation models hinge on exactly how parameter values are assigned (Stainforth et al., 2007).

the overall model (including other parameterizations). I leave aside that models of atmosphere and oceans get coupled, which arguably aggravates the problem.

Tuning is inevitable and is part and parcel of simulation modeling methodology. It poses great challenges, such as finding a good parameterization scheme for cloud dynamics, which is a recent area of intense research in meteorology. But when is a parameterization scheme a "good" one? On the one hand, a scheme is sound when it is theoretically well founded; on the other, the key property of a parameterization scheme is its adaptability. Both criteria do not point in the same direction. There is, therefore, no optimum: finding a balance is still considered an "art." I suspect that the widespread reluctance to publish articles on practices of adjusting parameters comes from reservations about addressing aspects that call for experience and art rather than theory and rigorous methodology.

Nothing in this argument is specific to climate. Climate modeling is just one example of many. The point holds for simulation modeling quite generally. Admittedly, climate might be a somewhat particular case, because it is placed in a political context in which some discussions seem to require admitting only ingredients of proven physical justification and realistic interpretation. Arguably, this expectation might motivate use of the pejorative term of "tuning." This reservation, however, ignores the very methodology of simulation modeling. Adjusting parameters is unavoidable and virtually ubiquitous in simulation.

Another example comes from thermodynamics, an area of physics with a very sound theoretical reputation. This example already appeared in chapter 4, where it illustrated the combinatorial style of modeling. The ideal gas equation is even taught in schools: It is an equation of state (EoS), describing how pressure and temperature depend on each other. However, more complicated equations of state find wide applications in chemical engineering as well. They are typically very specific for certain substances and require extensive adjustment of parameters—as Hasse and Lenhard (2017) have described and analyzed.

Clearly, being able to process specific adjustment strategies based on parameterization schemes is a crucial condition for success. Simulation methods have made thermodynamics applicable in many areas of practical relevance precisely because equations of state are tailored to particular cases of interest via the adjustment of parameters.

One further example comes from quantum chemistry—namely, density-functional theory (DFT), a theory developed in the 1960s for which Walter Kohn shared the 1998 Nobel Prize in chemistry. Density functionals capture the information from the Schrödinger equation, but are much more computationally tractable. However, only many-parameter functionals brought success in chemistry. The more tractable functionals with few parameters worked only in simpler cases of crystallography, but were unable to yield predictions of chemical interest. Arguably, being able to include and adjust more parameters has been the crucial condition that had to be satisfied before DFT could gain traction in computational quantum chemistry—which happened around 1990. This traction, however, is truly impressive. DFT is now the most widely used theory in scientific practice (see Lenhard, 2014, for a more detailed account of DFT and the development of computational chemistry).

Whereas the adjustment of parameters—to use the more neutral terminology—is pivotal for matching given data—that is, for predictive success—this very success condition also entails a serious disadvantage. Complicated schemes of adjusted parameters might block theoretical progress. In our climate case, any new cloud parameterization that intends to work with a more thorough theoretical understanding has to be developed over many years and then has to compete with a well-tuned forerunner. Again, this kind of problem is more general. In quantum chemistry, many-parameter adaptations of density functionals have brought great predictive success, but at the same time, this makes it difficult if not impossible to rationally reconstruct why such success occurs (Perdew et al., 2005; discussed in Lenhard, 2014). The situation in thermodynamics is similar (see Hasse & Lenhard, 2017).

Let me take stock regarding the first argument for the erosion of modularity. Tuning or adjusting parameters is not merely an ad hoc procedure to smoothen a model; rather, it is a pivotal component for simulation modeling. Tuning convolutes heterogeneous parts that do not have a common theoretical basis. Tuning proceeds holistically on the basis of global model behavior. How particular parts function often remains opaque. By interweaving local and global considerations, and by convoluting the interdependence of various parameter choices, tuning undermines modularity.

Looking back at Simon's clockmaker parable, one can see that its basic setting does not match the situation in a fundamental way. The perfect cogwheel picture is misleading, because it presupposes a clear identification of mechanisms and their interactions. In my examples, I showed that building a simulation model, unlike building a clockwork mechanism, cannot proceed in a top-down way. Moreover, different modules and their interfaces become convoluted during the processes of mutual adaptation.

EROSION OF MODULARITY 2: KLUGING

The second argument over the erosion of modularity approaches the issue from a different angle—namely, from a specific practice in developing software known as kluging.[14] *Kluge* is an invented English word from the 1960s that found its way into computer talk. I remember back in my childhood when we and another family drove off on vacation in two different cars. While crossing the Alps in the middle of the night, our friends' exhaust pipe broke off in front of us, creating a shower of sparks when the metal met the asphalt. There was no chance of getting the exhaust pipe repaired, but our friends' father did not hesitate long and used his necktie to fix it provisionally.

The necktie worked as a kluge, which is, in the words of *Wikipedia*, "a workaround or quick-and-dirty solution that is clumsy, inelegant, difficult to extend and hard to maintain." The notion has become popular in the language of software programming. It is closely related to the notion of bricolage. Andy Clark, for instance, stresses the important role played by kluges in complex computer modeling. For him, a kluge is "an inelegant, 'botched together' piece of program; something functional but somehow messy and unsatisfying." It is—Clark refers to Sloman—"a

14. Both spellings are used. There is not even agreement on how to pronounce the word. In a way, that fits the concept. I shall use *kluge* here, except when quoting authors who have used *kludge*.

piece of program or machinery which works up to a point but is very complex, unprincipled in its design, ill-understood, hard to prove complete or sound and therefore having unknown limitations, and hard to maintain or extend" (Clark, 1987, p. 278).

Kluge entered the body of philosophy guided by scholars such as Clark and Wimsatt, who are inspired by both computer modeling and evolutionary theory. The important point here is that kluges typically function for a whole system—that is, for the performance of the entire simulation model. In contrast, they have no meaning in relation to the submodels and modules: "What is a kludge considered as an item designed to fulfill a certain role in a large system may be no kludge at all when viewed as an item designed to fulfill a somewhat different role in a smaller system" (Clark, 1987, p. 279).

Because kluging is a colloquial term and is also not viewed as good practice, examples are difficult to find in published scientific literature. This observation notwithstanding, kluging is a widespread phenomenon. Let me give an example I encountered in visiting an engineering laboratory. There, researchers (chemical process engineers) were working with simulation models of an absorption column, the large steel structures in which reactions take place under controlled conditions. The scientific details do not matter here, because the point is that the engineers built their model on the basis of a couple of already existing modules, including proprietary software that they integrated into their simulation without having access to the code. Moreover, it is common knowledge in the community that this (black-boxed) code is of poor quality. Because of programming errors and because of ill- maintained interfaces, using this software package requires modifications to the remaining code to balance out unknown bugs. These modifications are not there for any good theoretical reasons, albeit are there for good practical ones. They make the overall simulation run as expected (in known cases); and they allow working with existing software. Hence, the modifications are typical kluges.

Again, kluging occurs in virtually every site at which large software programs are built. Hence, simulation models are a prime instance—especially when the modeling steps performed by one group build on the results (models, software packages) of other groups. One common phenomenon is the increasing importance of "exception handling"—that is, of finding effective repairs when the software or the model performs in unanticipated and undesired ways in some cases. In this situation, the software might include a bug that is invisible (does not affect results) most of the time, but becomes effective under particular conditions. Often, extensive testing is needed to find out about unwanted behavior that occurs in rare and particular situations that are conceived of as "exceptions"—indicating that researchers do not aim at a major reconstruction but, rather, at a local repair to counteract this particular exception.

Presumably all readers who have ever contributed to a large software program are familiar with experiences of this kind. It is commonly accepted that the more comprehensive a piece of software gets, the more energy new releases will require in order to handle exceptions. Operating systems of computers, for example, often receive weekly patches (i.e., repair updates). Many scientists who work with simulations face a similar situation, though not obviously so.

If, for instance, meteorologists want to work on, say, hurricanes, they are likely to take a mesoscale (multipurpose) atmospheric model from the shelf of some trusted modeling center and add specifications and parameterizations that are relevant for hurricanes. Typically, they will not know in exactly what respects the model had been tuned, and they will also lack much other knowledge about this particular model's strengths and weaknesses. Consequently, when preparing their hurricane modules, they will add measures to their new modules that somehow balance out any undesired model behavior. These measures can also be conceived as kluges.

Why should these examples be seen as typical instances and not as exceptions? Because they arise from the practical circumstances of developing software, and this is a core part of simulation modeling. Software engineering is a field that was envisioned as the "professional" answer to the increasing complexity of software. And I frankly admit that there are well-articulated concepts that would, in principle, ensure that software is clearly written, aptly modularized, well maintained, and superbly documented. However, the problem is that science *in principle* differs from science *in practice*.

In practice, strong and constant forces drive software developers to resort to kluges. Economic considerations are always a reason—be it on the personal scale of research time, or on the grand scale of assigning teams of developers to certain tasks. Usually, software is developed "on the move"; that is, those who write it have to keep up with changing requirements and meet a short deadline in both science and industry. Of course, in the ideal case, an implementation is tightly modularized. A virtue of modularity is that it is much quicker to incorporate "foreign" modules than to develop an entire enlarged program from scratch.

However, if these modules prove to have some deficiencies, developers will usually not start a fundamental analysis of how unexpected deviations occurred but, rather, invest their energy in adapting the interfaces so that the joint model will work as anticipated in the given circumstances. In common language, this is repair rather than replace. Examples extend from integrating a module of atmospheric chemistry into an existing general circulation model, to implementing the latest version of the operating system on your computer. Working with complex computational and simulation models seems to require a certain division of labor, and this division, in turn, thrives on software being able to travel easily. At the same time, this encourages kluges on the part of those who are trying to connect the software modules.

Kluges thus arise from unprincipled reasons: throwaway code, made for the moment, is not replaced later but instead becomes forgotten, buried in more code, and eventually is fixed. This will lead to a cascade of kluges. Once there, they prompt more kluges, thereby tending to become layered and entrenched.

Foote and Yoder, prominent leaders in the field of software development, give an ironic and funny account of how attempts to maintain a rationally designed software architecture constantly fail in practice.

> While much attention has been focused on high-level software architectural patterns, what is, in effect, the de facto standard software architecture is seldom discussed. This paper examines this most frequently deployed of software architectures: the BIG BALL OF MUD. A big ball of mud is a casually, even haphazardly, structured system. Its organization, if one can call it that, is dictated more by expediency than design. Yet, its enduring popularity cannot merely be

> indicative of a general disregard for architecture. . . . Reason for degeneration: ongoing evolutionary pressure, piecemeal growth: Even systems with well-defined architectures are prone to structural erosion. The relentless onslaught of changing requirements that any successful system attracts can gradually undermine its structure. Systems that were once tidy become overgrown as piecemeal growth gradually allows elements of the system to sprawl in an uncontrolled fashion. (1999, chap. 29)

I would like to repeat the statement from earlier that there is no necessity for the corruption of modularity and rational architecture. Again, this is a question of science in practice versus science in principle. "A sustained commitment to refactoring can keep a system from subsiding into a big ball of mud," Foote and Yoder admit. There are even directions in software engineering that try to counteract the degradation into Foote and Yoder's big ball of mud. The "clean code" movement, for instance, is directed to counter what Foote and Yoder describe. Robert Martin (2009), the pioneer of this school, proposes keeping code clean in the sense of not letting the first kluge slip in. And surely there is no principled reason why one should not be able to do this. However, even Martin is obliged to accept the diagnosis of current practice.

Similarly, Richard Gabriel (1996), another guru of software engineering, makes the analogy to housing architecture and Alexander's concept of "habitability" that aims to integrate modularity and piecemeal growth into one "organic order." In any case, when describing the starting point, he more or less duplicates what I mentioned earlier from Foote and Yoder.

Finally, I want to point out that the nature of kluging resembles what I introduced earlier under the notion of "opacity." Highly kluged software becomes opaque. One can hardly disentangle the various reasons that led to particular pieces of code, because kluges are meaningful only in one particular context, at one particular time. In this important sense, simulation models are historical objects. They carry around—and depend on—their history of modifications. There are interesting analogies with biological evolution that became a topic when complex systems developed into a major issue in the discussion on computer use.[15] Winograd and Flores (1991), for instance, come to a conclusion that also holds in our context here: "Each detail may be the result of an evolved compromise between many conflicting demands. At times, the only explanation for the system's current form may be the appeal to this history of modification" (p. 94).[16]

Thus, this brief look at the somewhat elusive field of software development has shown that two conditions foster kluging. The first is the exchange of software parts that is more or less motivated by flexibility and economic requirements. This thrives on networked infrastructure. The second is the need for iterations and modifications that are easy and cheap to perform. Owing to the unprincipled nature of kluges, their construction requires repeated testing for whether they actually work in the factual circumstances. Hence, kluges fit the exploratory and iterative mode of modeling that characterizes simulations. Furthermore, layered kluges solidify themselves. They make code hard or impossible to understand; modifying pieces that are individually hard to understand will normally lead to a new layer of kluges; and so on. Thus, kluging makes modularity erode, and this is the second argument as to why simulation modeling systematically undermines modularity.

THE LIMITS OF VALIDATION

What does the erosion of modularity mean for the validation of computer simulations? The key point regarding methodology is that holism is promoted by the very procedures that make simulation so widely applicable. But holism and the erosion of modularity are two sides of the same coin. It is through adjustable parameters that simulation models can be applied to systems beyond the control of theory (alone). The dissemination

15. Wimsatt (2007) writes about "generative entrenchment" when speaking about the analogy between software development and biological evolution; see also Lenhard and Winsberg (2010).
16. Interestingly, Jacob (1994) gives a very similar account of biological evolution when he writes that simpler objects are more dependent on (physical) constraints than on history, whereas history plays the greater part when complexity increases.

of simulation models and their swift adaptation to changing contexts promote practices of kluging. It is through these strategies that modularity erodes.

One ramification of utmost importance concerns the concept of validation. In the context of simulation models, the community speaks of *verification* and *validation*, or "V&V." Whereas verification checks the model internally (i.e., whether the software indeed captures what it is supposed to capture), validation checks whether the model adequately represents the target system. A standard definition states that "verification (is) the process of determining that a model implementation accurately represents the developer's conceptual description of the model and the solution to the model." In contrast, validation is defined as "the process of determining the degree to which a model is an accurate representation of the real world from the perspective of the intended uses of the model" (Oberkampf & Trucano, 2000, p. 3). Though there is some leeway in defining V&V, the gist of it is given in the saying "Verification checks whether the model is right, whereas validation checks whether you have the right model."

Owing to the increasing usage and growing complexity of simulations, the issue of V&V is itself an expanding topic in the simulation literature. One example is the voluminous monograph by Oberkampf and Roy (2010) that meticulously defines and discusses the various steps to be included in V&V procedures. A first move in this analysis is to separate model form from model parameters. Each parameter then belongs to a particular type of parameter that determines which specific steps in V&V are required. Oberkampf and Roy (2010, p. 623) give the following list of model parameter types:

- Measurable properties of the system or the surroundings
- Physical modeling parameters
- Ad hoc model parameters
- Numerical algorithm parameters
- Decision parameters
- Uncertainty modeling parameters

The adjustable parameters I discussed earlier do not appear in this list. This is not simply a matter of terminology. My point is that these parameters belong to both model form and model performance at the same time. The performance part of my claim is clear, because adjusting these parameters is oriented toward model performance. The crucial point is that these parameters also belong to the model form, because without the assignment of parameters, neither the question about representational adequacy nor the question about behavioral fit can be addressed. A cloud parameterization scheme makes sense only if its parameter values have been assigned already, and the same holds for an equation of state or for a many-parameter density functional. Before the process of adjustment, the mere form of the scheme can hardly be called adequate or inadequate. As I have shown, in simulation models, (predictive) success and adjustment are entangled. The adjustable parameters I have discussed are of a type that evades the V&V fencing.

Of course, the separation of verification and validation remains a conceptually valid and important distinction. However, it cannot be fully maintained in practice because it is not possible to first verify that a simulation model is "right" before starting to tackle the "external" question of whether it is the right model. I have touched upon this fact in chapter 3, on plasticity: The unspecified structure of a simulation model is often not sufficiently informative, whereas the full specification already involves adjusting parameters according to "external"—fit to data or phenomena—considerations. Performance tests hence become the main instrument for evaluating and confirming simulation models. This is a version of confirmation holism that points toward the limits of analysis. Again, this does not lead to a conceptual breakdown of verification and validation. Rather, holism comes in degrees[17] and is a pernicious tendency (*not* necessity) that undermines the verification–validation divide.[18]

Finally, I come back to the analogy, or rather, the disanalogy, between the computer and clockwork. In an important sense, computers

17. I thank Rob Moir for pointing this out to me.

18. My conclusion about the inseparability of verification and validation agrees well with Winsberg's (2010) more specialized claim in which he discusses model versions that evolve owing to changing parameterizations—a process that has been criticized by Morrison (2014). As far as I can see, her arguments do not apply to the case made here in this chapter that rests on a tendency toward holism, rather than a complete conceptual breakdown.

are not amplifiers; that is, they are not analogous to gigantic clockwork mechanisms. They do not (simply) amplify the force of mathematical modeling that has got stuck in overdemanding operations. Rather, computer simulation is profoundly *changing* the way mathematics is used. Throughout the book, I have analyzed simulation modeling as a new kind of mathematical modeling.

In the present chapter, I have questioned the rational picture of design. Brooks also did this when he observed that Pahl and Beitz had to include more and more steps to somehow capture an unwilling and complex practice of design, or when he referred to Donald Schön, who criticized a one-sided "technical rationality" underlying the "rational model" (Brooks, 2010, chap. 2). However, my criticism works, if you want, from "within." It is the very methodology of simulation modeling and how it works in practice that challenges the rational picture by making modularity erode.

The challenge to the rational picture is of a quite fundamental philosophical nature, because this picture has influenced how we conceptualize our world in so many ways. Many of these influences have not yet been explored. Just let me mention the philosophy of mind as one example. How we are inclined to think about mind today is influenced deeply by the computer and, in a more fundamental step, by our concept of mathematical modeling. Jerry Fodor (1983) has defended a most influential thesis that the mind is composed of information-processing devices that operate largely separately. Consequently, rethinking how computer models are related to modularity invites us to rethink the computational theory of the mind.

PART III

CONCLUSION AND OUTLOOK

8

Novelty and Reality

Chapters 1 through 7 presented an investigation of the philosophical characterization of simulation. This investigation confirmed that simulation modeling is a new type of mathematical modeling. This last chapter summarizes the findings, then presents an outlook on what I consider to be critical challenges for a philosophy of simulation. One of these challenges is to take into account the science–technology nexus. Another is to account for the relationship between human activity and reality that results from this nexus. This makes it necessary to rethink the instrumentalism versus realism divide in philosophy of science. In my opinion, simulation evades the stalemate of this divide. Finally, I gather together remarks about scientific rationality that are scattered across the preceding parts of this book. This leads me to conclude that simulation modeling has the potential to transform modern scientific rationality in a direction that looks surprisingly well known.

THE NOVEL AND THE USUAL

From the view of philosophy, what is probably the most frequent question about simulation asks whether it is new: yes or no? This suggests a black or white schema: either there has been a revolution or nothing new has happened. Right from the start, the present investigation has tried to qualify and undermine this somewhat crude dichotomy.

The answer of no has quite a lot to offer, because simulation does not present an entirely novel path for gaining knowledge. With this answer, it is subsumed under the more general category of mathematical modeling. But computer simulation offers genuine novelty *exactly because* it transforms the conception of mathematical modeling that has been (and still is) so fundamental to our understanding of what constitutes scientific rationality. The examples in the preceding chapters have shown that simulation modeling is a new type of mathematical modeling. The novelty is, as in the case of gunpowder, more a matter of the preparation than the ingredients. Important characteristics have been discussed and evaluated in chapters 1 through 4—namely, experimentation, artificial components, visualization, plasticity of models, and their epistemic opacity. These characteristics are interconnected and, in line with the summary in chapter 5, merge in an iterative and exploratory mode of modeling that is typical of simulation. In this mode, scientific ways of proceeding draw close to engineering ones. In particular, I have shown how this mode highlights the capacity for prediction while simultaneously questioning the explanatory power of simulation models. Chapters 6 and 7 scrutinized the seminal conceptions of solution and validation, pointing out how the transformation of mathematical modeling impinges on these conceptions. This delivers evidence for the claim that simulation modeling is not only a particular and new mode of mathematical modeling but also a style of reasoning in the sense of Ian Hacking (1992), who proposes that such a style impacts epistemology deeply and first determines what counts as right or wrong. What could have a more fundamental impact than the transformation of what is meant by (mathematical) solution, validation, or understanding?

This book was based on the decision to take mathematical modeling as a counterfoil for simulation modeling. I frankly admit that this decision is not without alternatives. I could, for example, have compared simulations to "real world experiments"[1] or to scale models. I did not attempt to characterize and analyze computer simulation in every respect.

Such an undertaking would involve struggling with the manifold purposes and methods of simulation studies. This investigation claims

1. Wolfgang Krohn (2007) has introduced this concept into the sociology of science. The same journal issue as contains his article provides a broad discussion on this topic.

to present *a* philosophy of computer simulation, not *the* philosophy. However, I would like to maintain that, as a guiding perspective for comparison, mathematical modeling has a special quality through being a centerpiece of modern science. As a result, theses on a new type of mathematical modeling can claim a general relevance for the philosophy of science.[2]

Characterizing simulation is a particularly difficult task, because the explicandum itself is a moving target. Simulation, computer-based modeling, and, in general, the ways computers are applied in the sciences are in an ongoing process of change. In a certain sense, the process can also be described as one of ongoing mathematization in new (as well as in established) fields that now increasingly implement computational approaches and models. At the outset, one can judge this from the sprawling growth of subdisciplines that carry "computational" in their name.[3] This new form of mathematization is working with the type of mathematical modeling investigated here; that is, it occurs typically in an exploratory and iterative mode.

Simulations permeate in so many ways into such a variety of sciences. Indeed, I am sure there are many instances in which they are taking pathways that cannot be classified yet as either right or wrong. Think of the tumultuous course of biological evolution when a new environment forms. Simulation procedures are still expanding rapidly, and any such process will probably produce a number of unsuccessful examples. A critical philosophical perspective, however, can contribute to comparing different simulation approaches and the degrees to which they are justified. Will it be possible to utilize the ultra-large plasticity of artificial neural networks in a broad range of applications?[4] Where do agent-based models, which

2. I think philosophers of mathematics such as Imre Lakatos would be amused, because, thanks to computational modeling, the thesis on the quasi-empirical character of mathematics, which Lakatos had based on a small number of historical cases, has now attained a broad footing in scientific practice.

3. For a recent account of this (partly) new wave of mathematization, see *Mathematics as a Tool* (Lenhard & Carrier, 2017).

4. During the time that this manuscript was being revised, "deep learning" algorithms in data-driven science have come to be advertised as an affirmative answer to this question.

are currently occupying largely unoccupied territory, deliver results of lasting significance? And so forth.

The close relationship between simulation and technology deserves extra attention in this context. This relationship can be recognized in the epistemology and methodology of simulation. One example is visualization. Technologies that display model output data immediately before the eyes of the modeler enable a quite direct interaction throughout the course of the modeling process. Prior to the 1980s, such technologies were not available, and consequently, visualization played a much smaller role in earlier phases of simulation modeling. Reversing this argument also seems plausible; that is, new technologies might lead to significant changes in the types of models that researchers prefer.

Up to now, digital electronic computers have been based on the Turing machine—not in terms of the technologies used in the circuitry but in terms of their logic and architecture—that is, what is called their "von Neumann architecture."[5] It has been the further development of computer technology that has strengthened the exploratory mode of modeling. One example for the impetus brought about by technological change is the role of the relatively small, networked computers that have become available in every office and laboratory. Today, desktop computers, workstations, or clusters of them are in use virtually everywhere in the sciences. They present a kind of alternative role model compared to centrally maintained computing centers, and they have strengthened the exploratory mode even further, because trying out and modifying a model comes at no (extra) cost in terms of researchers' money and energy. Computational modeling approaches have thrived with this networked infrastructure in a broad variety of sciences since the 1990s.[6]

5. It was John von Neumann's famous 1945 report on the EDVAC Computer that established this architecture (von Neumann, [1945]1981). One should note that the content of this report was essentially an outcome of collective achievements by a whole team of developers. Akera (2006) tells this story in a detailed and thoughtful manner. He points out how the work of Turing particularly influenced the conception of this architecture (see the 1946 report by Turing, 1986; for a broad argument, see Davis, 2000).

6. See Johnson and Lenhard (2011) for a programmatic exposition of this topic and Lenhard (2014) for a study on computational chemistry and the technology-related changes around 1990. Today, it appears as if data-intensive approaches so forcefully advertised by companies such as Google are reinforcing a regime of centrally controlled software and hardware.

CONSTRUCTING REALITY?

Thinking about how mathematical models become part of reality is of fundamental importance for epistemology. In this respect, the present investigation is inspired by Kant. When analyzing the nature of objective knowledge about the world, he emphasized the contribution of human constructive activity. This constructing involves empirical data, as well as theoretical concepts. The crucial point is that these ingredients develop a significant autonomous dynamic over the process of mediation. The world to be investigated and perceived is partly created in the process of knowledge production. Such a point of view brings together thinkers as diverse as Immanuel Kant, Ernst Cassirer, or Charles S. Peirce, along with Ian Hacking and some scholars in the philosophy of technology. One of the central issues in this line of philosophical thinking is to clarify how exactly construction (i.e., human activity) and reality or objective knowledge interrelate.[7]

From a bird's-eye perspective, this problem is about *the* characteristic of modern times. The Kantian impulse connects epistemology with technology, insofar as the latter also contributes to the dynamics of knowledge or, rather, influences how construction is practiced or even channeled. Seen from such a broad vantage point, the analyses of Paul Humphreys concluding that human subjects no longer own the center of epistemology (2004, p. 156) meet with those of Michel Foucault in *Les Mots et les Choses*. They are unlikely bedfellows, because Humphreys is dealing with the analytical philosophy of simulation, whereas Foucault is talking about the philosophical discourse of modernity. Of course, these differences are significant, and from a somewhat closer perspective, the analyses of both philosophers diverge significantly. Simulation modeling is a constructive activity, but this observation by itself does not characterize simulation. The point is, rather, *how* simulation *transforms* the classical conception of mathematical modeling.

7. In my opinion, "constructing reality" captures this relationship more aptly than "reconstructing reality"—as Morrison uses in her title (2014)—because the latter phrase assumes reality as something fixed in advance.

One motif recurs constantly in the literature on simulation in the philosophy of science: that simulation is making the distance between world and model disappear.

According to this position, computer-generated models and visualizations are emerging or functioning as a world of their own that is effectively replacing the real world. The concepts of virtual reality, or of a world *in silico*, capture much of this motif. It alludes to a counterfoil to traditional, presimulation modeling in which the models constructed by science worked as theoretical and idealized objects that had real-world counterparts (regardless of how complicated and indirect their relationship might be). Whereas the traditional models could be compared to the world—be it to explain certain phenomena or to intervene in them—this kind of comparison allegedly disappears with simulation, and the model world becomes self-sufficient or autarchic. A broad spectrum of thinkers supports some kind of self-sufficiency thesis, although this support does not exclude massive differences between these thinkers in other respects.

Peter Galison speaks about "computer as nature" (1996)—that is, about the tendency for the dynamics taking place in the computer to be regarded as the subject under investigation, hence replacing nature as the primary subject. Galison's position and similar ones face the general objection asking why a self-sufficient simulation should in any way work in applications. Galison addresses this objection by arguing that simulations are effective when applied to nature, because there is a deep ontological correspondence between computer and world. In particular, he takes stochasticity to be the decisive common property of both quantum processes in the world and Monte Carlo computer simulations. However, the material Galison provides to make a probabilistic world plausibly analogous to probabilistic simulations is less than convincing, because this material focuses on a rather narrow (and I also find somewhat exotic) segment of the history of Monte Carlo methods.

Stephen Wolfram advocates a similar thesis on simulation, substituting the world by drawing on an ontological argument. He claims that the common feature shared by both the world and cellular automata is their fractal geometry. This shared feature is the basis for his far-reaching claims about *A New Kind of Science* (2002). However, my investigation of simulation modeling has emphasized the plasticity of simulation models as one important feature. Against this background, it does not seem very plausible to adduce ontological reasons for why simulated phenomena fit their targets.

Alfred Nordmann (2006) follows a related line of thinking when he proposes that technoscience is characterized by a "collapse of distance." He argues that the distance between scientific representation and what is represented had always been the signature of modern science. Simulations then appear as examples of a broader movement away from classical modern science with its emphasis on the distance between model and world.

Moreover, simulations exemplify what Nordmann calls the shift from "science" to "technoscience." His point of view is in line with much of my foregoing analysis of simulation modeling. In particular, as I have shown, the complexity of the models forces one to regard them as objects that are partly constructed and investigated in their own right. In classical mathematical modeling, the distance between model and world had been marked by idealization, simplicity, and transparency. These properties have indeed become inadequate for describing simulation models. In other words, these properties no longer discriminate between the (simulation) model and the world.

I would like to mention Sherry Turkle as another person who can be viewed as belonging to this group. She highlights typical features of simulation from a sociological and ethnographic point of view (2009). Her main claim is that simulation involves a particular logic of action—namely, that researchers have to step into the model world: "simulation needs immersion." Researchers have no other choice, according to Turkle, than to immerse themselves in simulated worlds. This observation leads immediately to a worry, a worry Nordmann shares: How far are the acting scientists still aware of the fact that the (simulated) world they work in when carrying out their design studies and experiments is (merely) a constructed one? Turkle recognizes that a change of culture took place between 1980 and 2000. Whereas in 1980, simulation was new in many areas and was typically criticized against the backdrop of established practices—and therefore seen from a distance—one generation later, computer simulation has become an everyday practice, rarely questioned, and often without alternative.

In one respect, all the authors mentioned here, notwithstanding their differences, move toward a common denominator: the fundamental break linked to simulation allegedly consists in dropping the relationship to a world external to simulation. Nonetheless, let it be understood that the simulated world is not an arbitrary product of fantasy but, rather, is constructed and built according to a methodology aiming to find a match for certain dynamic properties. Therefore, I do not endorse this without reservation, because any application of this thesis to science and its uses of simulation has to be more nuanced. However, there are, in particular, two components of simulation modeling that deliver reasons favoring the autarky thesis—namely, the great complexity of the models and the exploratory mode of modeling connected to this complexity.

The increasing complexity of simulation models is associated with a growing division of labor. Teams of software developers build modules of ever larger software packages that are then used by researchers. Or, scientists in one discipline create a model that feeds into a larger simulation in a different discipline—to name just two typical instances. In many practical cases, the actual users of some code are not able to analyze it down to its roots; that is, they do not know the algorithmic mechanisms. It does not matter here whether this happens for an epistemological or an economical reason—because of the sheer size of the code or because the code is proprietary. Whatever the case, this is a sort of black-boxing that makes it hard to distinguish in what ways simulation results depend on certain assumptions and conditions. I discussed this observation in chapter 4 and also when addressing the limits of analysis in chapter 7.

At the same time, the number of scientists who use a computer to work on internal problems of computational modeling is increasing rapidly. "Internal" here means that work on the problems, as well as their solutions, is evaluated according to the simulation alone, as when a model has to be adapted to a new grid structure, a set of models needs to be made compatible (ready for coupling), or one parameterization scheme is to be replaced by a more efficient one. In other words, the increasing complexity of simulation models and the increasing complexity of the networked infrastructure deployed in the process of building and adapting these models are accompanied by an increase in the relative proportion of scientists working only on tasks that are defined and processed as being *internal* to simulation. This makes it a fact of social organization, and an important one, that models form a growing part of the modelers' world of experience.

A second reason favoring the autarky thesis comes from the observation that the methodology of simulation influences how models are perceived. The iterative and exploratory mode requires researchers to be very interactive—for example, when they modify models in response to the behavior they visualize. Hence, over the course of the modeling process, the modeled objects apparently exhibit resistance as if they belonged to a world external to the model. This mode promotes a certain potential for modelers to delude themselves regarding the constructed and mathematical nature of the simulated objects. For methodological reasons, there is no escape from intensively varying simulations interactively, iteratively, and often over a lengthy time span. The intensity required, together with the often-noted vivid appeal of visualizations, provides the second reason for the autarky stance.

In short, the organization of modeling, which is necessarily based on a strong division of labor, together with the way researchers interact with the simulation models, really does suggest the autarky thesis. However, suggesting is far from justifying. As a sociological observation, the thesis is valid—as illustrated by, for instance, Turkle's study. Nevertheless, this does not answer the philosophical and epistemological problem. In my opinion, the observations on the social and pragmatic level cannot support this far-reaching epistemological claim.

The analysis of simulation modeling also provides reasons to support a critical position—not critical in the sense of being negative but, rather, in the sense of being reflective. It is exactly the mode of simulation modeling expounded here that contains components counteracting the autarky thesis. The iterative and exploratory mode turns out to be a pragmatic means for circumventing—or trying to circumvent—the complexity barriers that impede mathematical modeling. One essential part of simulation modeling is to test the fit before modifying the model and starting the next iteration. In the course of such a procedure, the model maintains a preliminary status that points toward the next comparison—even if it is not clear whether this comparison points to something outside the model world.

Additionally, in many of the cases investigated here, modeling targeted the similarity between phenomena and how the models perform. To the extent that a model is plastic and works with artificial components, its specification has to be based on criteria that are not rooted in the model itself. To the extent that modeling aims to attain a correspondence on the level of performance, the model is precisely not autarkic.

It follows from my investigation that simulation results are preliminary for methodological reasons. This insight does not (just) come from general epistemological considerations on definite knowledge being outside the reach of human capacities but also from reasons specific to simulations, because simulation modeling deliberately calculates results that will be modified, therefore making them preliminary. For the model, failure and modification are the rule; they have their systematic place. The potential for surprises is something that the methodology takes into account. A model rarely jumps into the confirmation-or-refutation mode but, rather, moves on in an exploratory mode. Specifying the next modification and sounding out how the model behavior is affected are often relatively "cheap." In a way, a decision is swiftly postponed in favor of an additional loop of modification.

Simulation works with *calculated surprises*. Many scientists who work with computer models and who apply the behavior of these models in their work are perfectly aware of this fact. I have argued that the potential for adjustment and calibration is one of the strengths of simulation models. To interpret this strength as a sign of self-sufficiency would be unjustified. In fact, the opposite would make more sense: Adaptation and calibration are oriented toward something, and this is a reminder that simulation models, in a structural sense, are not saturated or self-sufficient but, instead, call for a counterpart.

In simulation, modelers are actively involved in building and constructing a world. Although this world is a calculated one, it offers surprises, both within the simulation (when models do not behave as anticipated in line with the assumptions entered into them) and outside it (when simulations match—or do not match—their target systems in unforeseen ways). Simulation modeling therefore demands that researchers orient themselves toward this world and respond to these surprises. Because this demand is a regular part of simulation methodology, surprises are something calculable. In short, calculated surprises are part and parcel of simulations.

RETHINKING INSTRUMENTALISM

I have mentioned the topic of instrumentalism several times in the preceding chapters, but I have postponed addressing it to here. Instrumental components figure prominently in simulation modeling. Hence, some version of instrumentalism would seem to deserve a central place in the philosophy of simulation. This immediately raises the question of how (this adequate conception of) instrumentalism relates to realism. This is not only an important question but also one that suggests itself because the controversial debate between proponents of instrumentalism and those of realism dominated much of philosophy of science in the 1980s. Does the philosophy of simulation side with instrumentalism? A straightforward yes would be rash.

A short introduction to the controversy is required here. The details of the debate fill volumes and have led to a stalemate. A well-informed formulation is Anjan Chakravartty's (2017) entry on "scientific realism" in the *Stanford Encyclopedia of Philosophy*. Realism, he puts it, is a "positive epistemic attitude toward the content of our best theories and models, recommending belief in both observable and unobservable aspects of the world described by the sciences." Instrumentalism holds that terms for unobservables do not have a meaning all by themselves. The latter condition is open to various interpretations. A sympathetic reading takes instrumentalism as a pragmatism of sorts. Science accordingly does not represent "the" (preexisting and uniquely determined) structure of reality but, rather, results from human and societal interaction with nature. Pragmatists like John Dewey (1912) see themselves as instrumentalists in this sense. There are many philosophical accounts that juxtapose realism with instrumentalism (see Chakravartty, 2017, for references), and there is considerable leeway in how liberally one defines instrumentalism or realism.[8]

8. Chakravartty (2017), for example, recognizes similarities between realism and constructive empiricism à la van Fraassen (1980) when the latter adopts an agnostic (not positive)

A strong conviction of instrumentalists that runs counter to realism is that science will develop further. Hence, any realistic interpretation at one time will become questionable at a later time. Realists typically acknowledge this suspicion and admit that change will happen, but they argue that it will not fundamentally affect theoretically well-founded (though unobservable) entities. Simulation, however, escapes from this coordinate system because it works with relatively "weak" objects—that is, objects that are, to an important extent, codefined by modeling decisions, adjustments, and discretizations.

A second main field of controversy is empirical underdetermination. Instrumentalists take this as an argument indicating that theoretical entities cannot be so well founded after all. Realists may counter that explanatory power provides an extra criterion that can single out the right option from a variety of options that are possible in principle. In simulation, however, opacity is a major obstacle to explanatory potential.

Hence, simulation modeling undermines the longstanding catch-22 situation in which instrumentalist and realist positions trump each other. Then, simulation instrumentalism is opposed not so strictly to realism. On the contrary, it takes a realist position in a pragmatist sense. Simulation has theory *and* instrument on board—in contrast to seeing them as strongly opposed. In my view, this is the most fundamental aspect of the *combinatorial* style of reasoning. "Isn't this just a makeshift?" the stern realist might reply, "a compromise that trades real realism for something much weaker?" I admit this reply has something to it. But admitting imperfection itself sounds like a very realistic position. Simulation models explore territory largely inaccessible for traditional (precomputer) methods. It would be an unwarranted move to presuppose that the ideal types of traditional science will remain unchanged.

It is as if there are wonderful fruits of science growing high up in well-rooted majestic trees. For the initiated, they look like a perfect delight. Much of history and philosophy of science is oriented toward them. Viewing such fruits as the natural limiting point of all scientific

attitude toward unobservables. If one is willing to do without this attitude, fictionalism, as proposed by Winsberg (2010) for the epistemology of simulation, also seems to be compatible with realism.

knowledge seems to be a position that fails to capture the gist of simulation. The combinatorial style of reasoning is much more suited to exploring a more earthly and dappled environment in which the fruits may not be so perfect, but they are numerous. Simulation thus has ramifications regarding how we conceive science and scientific research, its goals, and its rationality.

This and the preceding section lead to the following two points. One involves epistemology. Simulation combines constructive activity, instrumentalism, and realism—and thus evades longstanding opposing positions in philosophy of science. The second one is about the adequate perspective for studying science: Simulation-based science utilizes a new version of mathematical modeling to enter new territory. To the extent that mathematization forms a constitutive part of modern scientific rationality, the philosophy of simulation faces the question whether and how the new type of mathematical modeling affects and potentially changes the conception of rationality.

AN OUTLOOK ON RATIONALITY

The question whether and how simulation is transforming rationality is a main—and probably controversial—topic for any philosophy of simulation. The one I have offered in this book is oriented toward the conception of mathematical modeling as a rich source of similarities and differences. This conception itself has undergone profound changes in the course of the developing sciences. It is tied not only to the emergence of modern science but also to modernity in a more comprehensive sense. The transformations that led to modernity are helpful points of reference when judging the (potential) relevance of simulation.

Learning about scientific rationality might be the strongest reason why investigating simulation and computer modeling is of general philosophical importance. Let me paint a rough picture with just a very few broad brushstrokes. What our modern culture considers to be rationality is related intimately to the broader historical and intellectual development of society. Scientific rationality did not emerge at once in early modern times, but developed as a longer process, often by opposing what were then established forms of rationality.[9] The critical impetus of the Enlightenment documents this. The tensions, battles, and transformations accompanying industrialization turned large parts of society upside down.

The concept of rationality remained a contested one. Is scientific rationality something of a blueprint for society in general? Many and different parties approved this standpoint enthusiastically, whereas others were deeply skeptical—think of positions as diverse as nineteenth-century German Romanticism, Husserl's (1970 [1936]) *The Crisis of European Sciences and Transcendental Phenomenology*, or the writings of the "Critical School," such as Habermas's *Technology and Science as "Ideology"* (1970). These days, however, rationality is often defined as being more or less identical to scientific rationality.

What, then, characterizes scientific rationality? An essential part of the answer— though surely not the only part—can arguably be found in the entanglement of science and mathematics. One could reconstruct a grand movement of mathematization that started in the seventeenth century, reached a first peak with Newton, led to a scientization of engineering and technology in the late nineteenth century, and so forth. The computer would then appear as the instrument that makes mathematization jump over (some, though not all of) the hurdles of complexity, thus invigorating mathematization. However, I have argued that this picture is not tenable and may even be misleading. Simulation modeling is not in line with this narrative of mathematization. Rather, it transforms mathematical modeling and has been described here as an iterative and exploratory mode of modeling and a combinatorial style of reasoning. The profound transformations of concepts such as solution, validation, and modularity question any narrative about science as straightforwardly unfolding rationality. It is the complexities and complications of simulation, rather than its algorithmic logic, that affect epistemology in the philosophically most interesting way.

For reasons of comparison, I find it instructive to look at figures positioned at the border to modernity. In the sixteenth-century late Renaissance period, one can see that the formation of modern rationality

9. Shapin and Shaffer's *Leviathan and the Air-Pump* (1985) is just one of the masterful accounts of this process that combine historical and philosophical perspectives.

had already started, and at this stage, it can be traced while it still was a very open process. Here is an illustration: Leonardo da Vinci (1452–1519) is widely recognized as a visionary, but there is controversy over whether he worked scientifically in any relevant sense of the word. In fact, one finds Leonardo classified as a physicist, a scientist, an artist, and even as surely *not* a scientist.[10] Whether one would count Leonardo among the scientists (*avant la lettre*) surely depends on which concept of science one adopts. One important aspect of this decision is the role mathematization supposedly plays in how the sciences develop. Leonardo was not at all interested in the contemporary academic mathematics that he arguably had not mastered well. He was very keen, however, on methods for treating practical problems in a quantitative manner, such as problems in statics.

> More significantly, Leonardo showed no interest whatever in the exciting development of algebra in his own age. What is more, in geometry he disregarded the traditional limitation of construction by straight edge and compass; he was quite content with numerical approximations. (Seeger, 1974, p. 45)

This unorthodox and anti-academic attitude nevertheless conceals a strong interest in mathematization. Leonardo oriented himself toward treating problems in a quantitative manner (e.g., through measurements instead of traditional theories) and toward ideas guided by technology and experience (according to the engineer Reti, cited in O'Malley, 1968).

Leonardo structured his activities as an instrument maker and artist by taking a mathematizing approach—not in the sense of theory but, rather, in the sense of quantification and mathematical representation. In this way, mathematization should support descriptive and technical processes and designs. Integrating series of exploratory experiments and tests, as Leonardo proclaimed, resembled approaches such as Tartaglia's "practical geometry" more than the formulation of a theory. Randall describes the same finding in different words: "Leonardo was congenial more to Edison than to Einstein" (Randall, 1961, pp. 131–132).

10. For a compilation of controversial opinions, see Seeger (1974).

Leonardo's philosophical diaries reveal repeated praise of experience versus dogma, supplemented consistently by expressing a quite bold acceptance that his works and results tended to be unfinished, imperfect, and even preliminary. This attitude surely was directed against the encyclopedism of his time and was bolstered by an awareness of the manifold and complicated character of the problems being tackled (see, e.g., Leonardo, 1958, p. 167). After all, one can distinguish something like a principle in Leonardo of basing his view of the world on experience, experiment, and mathematics—a stance directed just as much against alchemists as against theologians. In this regard, Leonardo appears surprisingly modern, even though he was not widely received as such, making his factual influence on the further development of the sciences rather limited.[11] Incorporating a mathematical perspective into a preliminary, explorative approach looks surprisingly similar to the combinatorial style of reasoning that characterizes simulation.

Here is a second illustration following my extremely eclectic choice. It is well known that Francis Bacon (1561–1626) took practical utility as a major touchstone for theoretical science. Even if science could attain some genuine understanding according to its own measures, this could not possibly replace a lack of "practical benefits" (Urbach, 1987, p. 14). Bacon basically envisioned a "symbiosis of science and technology" (Krohn, 2006, p. 192) in which scientific knowledge and capability for carrying out interventions would coincide.

> In a nutshell: if a scientific statement can lead to a successful (re) production of the phenomenon it purports to describe, however crudely or approximately, then that very statement should be accepted as a full denizen into the realm of practically sanctioned knowledge. (Pérez-Ramos. 1996, p. 111)

This statement resembles the motto of "saving the phenomena." Hence, simulation-based knowledge would have been much in line with Bacon's stance, because he considered a core issue of the sciences to be

11. Drake and Drabkin (1969) qualify the purportedly central role that Duhem reserved for Leonardo.

the establishment of the power to carry out interventions via technical constructions. This goal was so important that understanding appeared to be derived from it: "only the doable—at least in principle—is also understandable" (Funkenstein, 1986, p. 78). Such a standpoint presupposes a successful turning away from Aristotle, who considered *artificialia* to be inadequate for gaining insight into nature because they belonged to a different realm. In my analysis of simulation, a very non-Aristotelean artificiality has emerged as an important element. In his verdict "nature to be commanded must be obeyed," Francis Bacon expressed clearly that technology and nature are positioned on the same level. He projected science as being little different from the art of engineering. This conception is much more fitting to recent developments in simulation modeling than to the trajectory that modern science—and mathematization—happened to follow in the time after Bacon.

Stephen Toulmin (1990) provided a very instructive account on how modernity emerged as an escape from severe and cruel wars in the first half of the seventeenth century. Modern rationalism firmly established stability, simplicity, certainty, and systematicity as guiding principles. The earlier prewar Renaissance humanism, so Toulmin expounds, was characterized by a "commitment to intellectual modesty, uncertainty, and toleration" (p. 174). Simulation modeling, in a seemingly paradoxical move, hence introduces notions that Toulmin had used to characterize sixteenth-century humanism. As a historian of rationality, Toulmin called for a combination of humanism and rationalism: "We are not compelled to choose between 16th-century humanism and 17th-century exact science: rather, we need to hang on to the positive achievements of them both" (p. 180).

I find it intriguing to place simulation modeling in this perspective. Simulation acknowledges that mathematical modeling need not hunt for underlying simplicity, stability, or uniformity, but accepts adaptability and diversity as additional guiding principles that sometimes take the lead. This suggests an ironic point: simulation modeling is surely a descendant in the line of modern rationality. Instead of fulfilling the latter's promises, however, it undermines them. The exploratory mode is acknowledging a "human" standpoint of tolerance, uncertainty, and modesty. At least it is an invitation to do so. That the present investigation of simulation leads to this kind of broad philosophical challenge came, I frankly admit, as an uncalculated surprise quite late when writing and revising this book.

Independent of whether one follows this historical-philosophical sketch with a positive attitude or a skeptical frown, so much seems to be clear: Simulation modeling is positioned in the midst of ongoing transformations that are happening on a local scale of science studies, as well as on the grand scale of philosophy. Coming to grips with them is a major challenge for any philosophy of simulation. As far as I can see, this challenge will not only provoke a thriving field of specialized philosophy (which it already does) but also address an important societal need for reflection. The very last part of the concluding chapter has now opened up a new challenge for the philosophy of science. If someone objects that a conclusion that opens up does not do justice to the very meaning of the word, I fully accept the criticism. Moreover, I welcome it.

REFERENCES

Agre, Philip E. 2003. "Hierarchy and History in Simon's 'Architecture of Complexity.'" *Journal of the Learning Sciences* 3: 413–426.

Ahrweiler, Petra, & Nigel Gilbert, eds. 1998. *Simulations in Science and Technology Studies*. Berlin: Springer.

Akera, Atsushi. 2006. *Calculating a Natural World: Scientists, Engineers, and Computers During the Rise of U.S. Cold War Research*. Cambridge, MA: MIT Press.

Anderson, Chris. 2008. "The End of Theory: The Data Deluge Makes the Scientific Method Obsolete." *Wired* Magazine, June.

Appel, Kenneth, & Wolfgang Haken. 1977. "The Solution of the Four Color Map Problem." *Scientific American* 137 (8): 108–121.

Appel, Kenneth, Wolfgang Haken, & John Koch. 1977. "Every Planar Map Is Four Colorable." *Illinois Journal of Mathematics* 21 (48): 429–567.

Arakawa, Akio. 1966. "Computational Design for Long-Term Numerical Integration of the Equations of Fluid Motion: Two-Dimensional Incompressible Flow. Part I." *Journal of Computational Physics* 1: 119–143.

Arakawa, Akio. 2000. "A Personal Perspective on the Early Years of General Circulation Modeling at UCLA." In *General Circulation Model Development*, edited by David A. Randall, 1–66. San Diego: Academic Press.

Ashby, William R. 1957. *An Introduction to Cybernetics*. 2nd ed. London: Chapman & Hall.

Aspray, William F. 1990. *John von Neumann and the Origins of Modern Computing*. Cambridge, MA: MIT Press.

Axelrod, Robert M. 1997. "Advancing the Art of Simulation." In *Simulating Social Phenomena*, edited by Rosaria Conte, Rainer Hegselmann, & Pietro Terno, 21–40. Berlin: Springer.

Bailer-Jones, Daniela. 2003. "When Scientific Models Represent." *International Studies in the Philosophy of Science* 17 (1): 59–74.

Baird, Davis, Alfred Nordmann, & Schummer Joachim, eds. 2004. *Discovering the Nanoscale*. Amsterdam: IOS Press.

Baker, Alan. 2008. "Experimental Mathematics." *Erkenntnis* 68 (3): 331–344.

Barberousse, Anouk, Sara Franceschelli, & Cyrille Imbert. 2007. "Cellular Automata, Modeling, and Computation." *Synthese* 169 (3): 557–574.

Bensaude-Vincent, Bernadette. 2001. "The Construction of a Discipline: Materials Science in the United States." *Historical Studies in the Physical and Biological Sciences* 31 (2): 223–248.

Bjerknes, Vilhelm. 1904. "Das Problem Der Wettervorhersage, Betrachtet Vom Standpunkte Der Mechanik Und Der Physik." *Meteorologische Zeitschrift* 21: 1–7.

Block, Ned, & Jerry Fodor. 1972. "What Psychological States Are Not." *Philosophical Review* 81: 159–181.

Blumenberg, Hans. 1965. "Das Fernrohr Und Die Ohnmacht Der Wahrheit." In *Galileo Galilei: Sidereus Nuncius. Nachricht von Neuen Sternen*, 7–75. Frankfurt: Suhrkamp.

Boden, Margaret, ed. 1996. *The Philosophy of Artificial Life*. Oxford: Oxford University Press.

Borwein, Jonathan M., & David H. Bailey. 2004. *Mathematics by Experiment: Plausible Reasoning for the 21st Century*. Natick, MA: AK Peters.

Borwein, Jonathan M., David H. Bailey, & Roland Girgensohn. 2004. *Experimentation in Mathematics: Computational Paths to Discovery*. Natick, MA: AK Peters.

Braitenberg, Valentin, & Inga Hosp, eds. 1995. *Simulation. Computer Zwischen Experiment und Theorie*. Reinbek bei Hamburg: Rowohlt.

Brandom, Robert. 2000. *Articulating Reasons: An Introduction to Inferentialism*. Cambridge, MA: Harvard University Press.

Bredekamp, Horst. 2003. "Bildwissenschaft." *Metzler Lexikon Kunstwissenschaft*. Stuttgart: Metzler.

Brennan, Robert D. 1968. "Simulation Is Wha-a-at?" In *Simulation. The Dynamic Modeling of Ideas and Systems with Computers*, edited by John McLeod, 5–12. New York: McGraw-Hill.

Brooks, Frederic P. 2010. *The Design of Design*. Boston, MA: Addison-Wesley.

Brooks, Rodney. 1991. "Intelligence Without Reason." A. I. Memo 1293. MIT.

Brooks, Rodney. 2002. *Flesh and Machines*. New York: Pantheon.

Brunt, David. 1944. *Physical and Dynamical Meteorology*. 2nd edition. Cambridge: Cambridge University Press.

Burtt, Edwin A. 1964. *The Metaphysical Foundations of Modern Physical Science: A Historical and Critical Essay*. London: Routledge & Kegan Paul.

Calude, Christian S., & Giuseppe Longo. 2017. "The Deluge of Spurious Correlations in Big Data." *Foundations of Science* 22 (3): 595–612.

Carrier, Martin. 2010. "Theories for Use: On the Bearing of Basic Science on Practical Problems." In *EPSA Epistemology and Methodology of Science: Launch of the European Philosophy of Science Association*, edited by Mauricio Suárez, Mauro Dorato, & Miklós Rédei, 23–34. Dordrecht: Springer.

Carrier, Martin, & Michael Stöltzner. 2007. "Modelle Als Bausteine Angewandter Wissenschaft." In *Nachrichten Aus Der Wissensgesellschaft*, edited by Peter Weingart, Martin Carrier, & Wolfgang Krohn, 164–181. Weilerswist: Velbrück.

Cartwright, Nancy. 1983. *How the Laws of Physics Lie*. Oxford & New York: Clarendon.

Cassirer, Ernst. 1910. *Substanzbegriff und Funktionsbegriff: Untersuchungen über die Grundfragen der Erkenntniskritik* [*Substance and Function*]. Berlin: Bruno Cassirer. [Reprint, Chicago: Open Court, 1923.]

Casti, John L. 1997. *Would-Be Worlds: How Simulation Is Changing the Frontiers of Science.* New York: John Wiley.

Ceruzzi, Paul E. 2003. *A History of Modern Computing.* Cambridge, MA, & London:: MIT Press.

Chaitin, Gregory J. 1998. *The Limits of Mathematics. A Course on Information Theory and the Limits of Formal Reasoning.* Berlin & New York: Springer.

Chakravartty, Anjan. 2017. "Scientific Realism." In *The Stanford Encyclopedia of Philosophy* (Summer 2017 Edition), edited by Edward N. Zalta. https://plato.stanford.edu/archives/sum2017/entries/scientific-realism/.

Chen, Min, & Luciano Floridi. 2013. "An Analysis of Information Visualisation." *Synthese* 190 (16): 3421–3438. https://doi.org/10.1007/s11229-012-0183-y.

Clark, Andy. 1987. "The Kludge in the Machine." *Mind and Language* 2 (4): 277–300.

Clark, Andy. 2003. *Natural-Born Cyborgs: Minds, Technologies, and the Future of Human Intelligence.* Oxford & New York: Oxford University Press.

Clark, Andy, & Josefa Toribio. 1994. "Doing Without Representing?" *Synthese* 101 (3): 401–431.

Conte, Rosaria, Rainer Hegselmann, & Pietro Terno, eds. 1997. *Simulating Social Phenomena*. Berlin: Springer.

Crombie, Alistair C. 1988. "Designed in the Mind: Western Visions of Science, Nature and Humankind." *History of Science* 26: 1–12.

Crombie, Alistair C. 1994. *Styles of Scientific Thinking in the European Traditions*, Vols. I–III. London: Duckworth.

Daston, Lorraine, & Peter Galison. 2007. *Objectivity*. New York: Zone Books.

Dear, Peter. 2006. *The Intelligibility of Nature. How Science Makes Sense of the World.* Chicago: University of Chicago Press.

DeMillo, Richard A., Richard J. Lipton, & Alan J. Perlis. 1977. "Social Processes and Proofs of Theorems and Programs." In *Proceedings of the 4th ACM Symposium on Principles of Programming Languages*, 206–214. New York: ACM.

de Regt, Henk W., & Dennis Dieks. 2005. "A Contextual Approach to Scientific Understanding." *Synthese* 144 (1): 137–170.

de Regt, Henk W., Sabina Leonelli, & Kai Eigner. 2009. *Scientific Understanding: Philosophical Perspectives*. Pittsburgh, PA: University of Pittsburgh Press.

Detlefsen, Michael. 1986. *Hilbert's Program: An Essay on Mathematical Instrumentalism*. Dordrecht: Reidel.

Diaconis, Persi. 2009. "The Markov Chain Monte Carlo Revolution." *Bulletin of the American Mathematical Society* 46: 179–205.

Diderot, Denis. [1754]1999. *Thoughts on the Interpretation of Nature*. Translated by Lorna Sandler. Manchester, UK: Clinamen.

Dijksterhuis, Eduard Jan. 1961. *The Mechanization of the World Picture*. London: Oxford University Press.

Di Paolo, Ezequiel A., Jason Noble, & Seth Bullock. 2000. "Simulation Models as Opaque Thought Experiments." In *Artificial Life VII: Proceedings of the Seventh International Conference on Artificial Life*, edited by Mark Bedau, J. S. McCaskill, N. H. Packard, & S. Rasmussen, 497–506. Cambridge, MA: MIT Press.

Dowling, Deborah. 1999. "Experimenting on Theories." *Science in Context* 12 (2): 261–273.

Drake, Stillman, & Israel E. Drabkin, eds. 1969. *Mechanics in Sixteenth-Century Italy. Selections from Tartaglia, Benedetti, Guido Ubaldo, and Galileo*. Madison, Milwaukee and London: The University of Wisconsin Press.

Dueck, Gunter. 1993. "New Optimization Heuristics: The Great Deluge Algorithm and the Record-to-Record Travel." *Journal of Computational Physics* 104 (1): 86–92.

Dueck, Gunter, Tobias Scheuer, & Hans-Martin Wallmeier. 1999. "Toleranzschwelle und Sintflut: Neue Ideen zur Optimierung." *Spektrum der Wissenschaft* 2: 22–31.

Eady, Edward. 1956. "Discussions." *Quarterly Journal of the Royal Meteorological Society* 82: 535–539.

Edwards, Paul N. 1996. *The Closed World: Computers and the Politics of Discourse in Cold War America*. Cambridge, MA: MIT Press.

Elman, J. 1994. "Learning and Development in Neural Networks: The Importance of Starting Small." *Cognition* 48: 71–99.

Epstein, Joshua M., & Robert L. Axtell. 1996. *Growing Artificial Societies. Social Science from the Bottom Up*. Washington, DC, & Cambridge, MA: Brookings Institution and MIT Press.

Evstigneev, Mykhailo, & Peter Reimann. 2003. "Dynamic Force Spectroscopy: Optimized Data Analysis." *Physical Review E* 68 (045103): 1–4.

Fetzer, James. 1988. "Program Verification: The Very Idea." *Communications of the ACM* 31: 1048–1063.

Feynman, Richard P. 1960. "There Is Plenty of Room at the Bottom." Talk published 1961 in *Miniaturization*, edited by Horace D. Gilbert, 282–296. New York: Reinhold.

Feynman, Richard P. 1985. *Surely You're Joking, Mr. Feynman! Adventures of a Curious Character.* Edited by Ralph Leighton. New York: W.W. Norton.

Feynman, Richard P., Robert B. Leighton, & Matthew Sands. 1965. *The Feynman Lectures on Physics.* Reading, MA: Addison-Wesley.

Fillion, Nicolas. 2017. "The Vindication of Computer Simulations." In *Mathematics as a Tool,* edited by Johannes Lenhard & Martin Carrier, 137–155. Boston Studies in the History and Philosophy of Science 327. New York: Springer.

Fillion, Nicolas, & Sorin Bangu. 2015. "Numerical Methods, Complexity, and Epistemic Hierarchies." *Philosophy of Science* 82: 941–955.

Fillion, Nicolas, & Robert M. Corless. 2014. "On the Epistemological Analysis of Modeling and Computational Error in the Mathematical Sciences." *Synthese* 191: 1451–1467.

Fodor, Jerry. 1983. *The Modularity of Mind.* Cambridge, MA: MIT Press.

Foer, Franklin. 2017. *World Without Mind: The Existential Threat of Big Tech.* New York: Penguin.

Foote, Brian, & Joseph Yoder. 1999. "Big Ball of Mud." In *Pattern Languages of Program Design 4,* edited by Neil Harrison, Brian Foote, & Hans Rohnert, 653–692. Boston, MA: Addison-Wesley. http://www.laputan.org/mud/mud.html.

Frenkel, Daan. 2013. "Simulations: The Dark Side." *European Physical Journal Plus* 128 (10). https://doi.org/DOI 10.1140/epjp/i2013-13010-3.

Frigg, Roman, & Stephan Hartmann. 2008. "Models in Science." In *The Stanford Encyclopedia of Philosophy,* edited by Edward N. Zalta. http://plato.stanford.edu/archives/fall2008/entries/models-science/.

Frigg, Roman, & Julian Reiss. 2009. "The Philosophy of Simulation. Hot New Issues or Same Old Stew?" *Synthese* 169 (3): 593–613.

Funkenstein, Amos. 1986. *Theology and the Scientific Imagination From the Middle Ages to the Seventeenth Century.* Princeton, NJ: Princeton University Press.

Gabriel, Richard P. 1996. *Patterns of Software: Tales From the Software Community.* New York & Oxford: Oxford University Press.

Galison, Peter. 1994. "The Ontology of the Enemy: Norbert Wiener and the Cybernetic Vision." *Critical Inquiry* 21: 228–266.

Galison, Peter. 1996. "Computer Simulations and the Trading Zone." In *The Disunity of Science: Boundaries, Contexts, and Power,* edited by Peter Galison & David J. Stump, 118–157. Stanford, CA: Stanford University Press.

Galison, Peter. 1997. *Image and Logic: A Material Culture of Microphysics.* Chicago & London: University of Chicago Press.

Galois, Évariste. 1962. "Sur la Méthode." In *Écrits et Mémoires Mathématiques d'Évariste Galois,* edited by Robert Bourgne & Jean-Pierre Azra, 2–11. Paris: Gauthier-Villars.

Gao, Jianping, Walker D. Luedtke, & Uzi Landman. 1998."Friction Control in Thin-Film Lubrication." *Journal of Physical Chemistry B* 102: 5033–5037.

Gardner, Martin. 1970. "Mathematical Games: The Fantastic Combinations of John Conway's New Solitaire Game 'Life.'" *Scientific American* 223: 120–123.

Gibbons, Michael, Camille Limoges, Helga Nowotny, Simon Schwartzman, Peter Scott, & Martin Trow. 1994. *The New Production of Knowledge: The Dynamics of Science and Research in Contemporary Sciences*. London: Sage.

Giere, Ron. 1988. *Explaining Science. A Cognitive Approach*. Chicago: University of Chicago Press.

Gooding, David, Trevor Pinch, & Simon Schaffer, eds. 1989. *The Uses of Experiment*. Cambridge: Cambridge University Press.

Gramelsberger, Gabriele. 2010. *Computerexperimente zum Wandel der Wissenschaft im Zeitalter des Computers*. Bielefeld: Transkript.

Gramelsberger, Gabriele. 2011. *From Science to Computational Sciences*. Zürich: Diaphanes.

Gramelsberger, Gabriele, & Johann Feichter, eds. 2011. *Climate Change and Policy: The Calculability of Climate Change and the Challenge of Uncertainty*. Heidelberg: Springer.

Grier, David A. 2005. *When Computers Were Human*. Princeton, NJ: Princeton University Press.

Grüne-Yanoff, Till. 2009. "The Explanatory Potential of Artificial Societies." *Synthese* 169 (3): 539–555.

Gunderson, Keith, ed. 1985a. *Mentality and Machines*. 2nd ed. Minneapolis: University of Minnesota Press.

Gunderson, Keith. 1985b. "Philosophy and Computer Simulation." In *Mentality and Machines*, 2nd ed., edited by Keith Gunderson, 91–135. Minneapolis: University of Minnesota Press.

Gutin, Gregory, & Abraham P. Punnen, eds. 2002. *The Traveling Salesman Problem and Its Variations*. Dordrecht: Kluwer.

Habermas, Jürgen. 1970. "Technology and Science as 'Ideology'". In *Towards a Rational Society*, 81–127. Boston: Beacon Press.

Hacking, Ian. 1983. *Representing and Intervening: Introductory Topics in the Philosophy of Natural Science*. Cambridge: Cambridge University Press.

Hacking, Ian. 1992. "'Style' for Historians and Philosophers." *Studies in the History and Philosophy of Science* 23 (1): 1–20.

Hartmann, Stephan. 1996. "The World as a Process: Simulations in the Natural and Social Sciences. In *Modelling and Simulation in the Social Sciences from the Philosophy of Science Point of View*, edited by Rainer Hegselmann, 77–100. Dordrecht: Kluwer.

Hasse, Hans, & Johannes Lenhard. 2017. "Boon and Bane: On the Role of Adjustable Parameters in Simulation Models." In *Mathematics as a Tool: Tracing New Roles of Mathematics in the Sciences*, edited by Johannes Lenhard & Martin Carrier, 93–115. Boston Studies in the Philosophy and History of Science 327. New York: Springer.

Hasslacher, Brosl. 1987. "Discrete Fluids." *Los Alamos Science* 15: 175–217.

Hegerl, Gabriele C., Klaus Hasselmann, U. Cubasch, J. F. B. Mitchell, E. Roeckner, R. Voss, & J. Waszkewitz . 1997. "Multi-Fingerprint Detection and Attribution Analysis of Greenhouse Gas, Greenhouse Gas-Plus-Aerosol and Solar Forced Climate Change." *Climate Dynamics* 13: 613–634.

Hegselmann, Rainer, ed. 1996. *Modelling and Simulation in the Social Sciences from the Philosophy of Science Point of View*. Dordrecht: Kluwer.

Heidelberger, Michael. 2006. "Applying Models in Fluid Dynamics." *International Studies in Philosophy of Science* 20 (1): 49–67.

Heidelberger, Michael, & Friedrich Steinle, eds. 1998. *Experimental Essays*. Baden-Baden: Nomos Verlagsgesellschaft.

Heims, Steve J. 1980. *John von Neumann and Norbert Wiener: From Mathematics to the Technologies of Life and Death*. Cambridge, MA, & London: MIT Press.

Heintz, Bettina, & Jörg Huber, eds. 2001. *Mit dem Auge Denken. Strategien der Sichtbarmachung in Wissenschaftlichen und Virtuellen Welten*. Vienna & New York: Springer.

Helbing, Dirk. 1995. *Quantitative Sociodynamics: Stochastic Methods and Models of Social Interaction Processes*. Dordrecht: Kluwer.

Hey, Tony, Stewart Tansley, & Kristin Tolle, eds. 2009. *The Fourth Paradigm: Data-Intensive Scientific Discovery*. Redmond, WA: Microsoft Research.

Heymann, Matthias, & Helge Kragh, eds. 2010. "Modelling and Simulation in the Atmospheric and Climate Sciences." *Studies in History and Philosophy of Science Part B: Studies in History and Philosophy of Modern Physics* 41 (3): 193–292.

Hofmann, Joseph E. 1974. *Leibniz in Paris 1672–1676: His Growth to Mathematical Maturity*. Cambridge: Cambridge University Press.

Holland, Olaf A. 1987. "Schnittebenenverfahren für Travelling-Salesman-und Verwandte Probleme." Ph.D. dissertation, Bonn University.

Houghton, John T., Y. Ding, D. J. Griggs, M. Noguer, P. J. van der Linden, & D. Xiaosu, eds. 2001. *Contribution of Working Group I to the Third Assessment Report of the Intergovernmental Panel on Climate Change (IPCC)*. Cambridge: Cambridge University Press.

Hourdin, Frederic, Jean-Yves Grandpeix, Catherine Rio, Sandrine Bony, Arnaud Jam, Frédérique Cheruy, Nicolas Rochetin, Laurent Fairhead, Abderrahmane Idelkadi, Ionela Musat, Jean-Louis Dufresne, alain Lahellec, Marie-Pierre Lefebvre, & Romain Roehrig. 2013. "LMDZ5B: The Atmospheric Component of the IPSL Climate Model with Revisited Parameterizations for Clouds and Convection." *Climate Dynamics* 40: 2193–2222.

Hubig, Christoph. 2008. "Virtualisierung der Technik—Virtualisierung der Lebenswelt? Neue Herausforderungen für eine Technikethik als Ermöglichungsethik." Talk presented at the XXI. Deutscher Kongress für Philosophie, Essen, Germany.

Hughes, R. I. G. 1997. "Models and Representation." *Philosophy of Science* 64: S325–336.

Hughes, R. I. G. 1999. "The Ising Model, Computer Simulation, and Universal Physics." In *Models as Mediators*, edited by Mary Morgan & Margaret Morrison, 97–145. Cambridge: Cambridge University Press.

Humphreys, Paul W. 1991. "Computer Simulations." In *PSA 1990*, edited by Arthur Fine, F. Forbes, & L. Wessels, 2:497–506. East Lansing, MI: Philosophy of Science Association.

Humphreys, Paul W. 1994. "Numerical Experimentation." In *Patrick Suppes: Scientific Philosopher*, edited by Paul W. Humphreys, 2:103–121. Dordrecht: Kluwer.

Humphreys, Paul W. 2004. *Extending Ourselves: Computational Science, Empiricism, and Scientific Method*. New York: Oxford University Press.

Humphreys, Paul W. 2009. "The Philosophical Novelty of Computer Simulation Methods." *Synthese* 169 (3): 615–626.

Humphreys, Paul W. 2012. "Computational Science." *Oxford Bibliographies*. Online. http://www.oxfordbibliographies.com/view/document/obo-9780195396577/obo- 9780195396577-0100.xml.

Husserl, Edmund. 1970. *The Crisis of European Sciences and Transcendental Phenomenology: An Introduction to Phenomenological Philosophy*. Originally published in German 1936. Evanston, IL: Northwestern University Press.

Ihde, Don. 2004. "Has the Philosophy of Technology Arrived?" *Philosophy of Science* 71: 117–131.

Ihde, Don. 2006. "Models, Models Everywhere." In *Simulation: Pragmatic Construction of Reality*, edited by Johannes Lenhard, Günther Küppers, & Terry Shinn, 79–86. Yearbook Sociology of the Sciences 25. Dordrecht: Springer.

Jacob, Francois. 1994. *The Possible and the Actual*. Seattle: University of Washington Press.

Jahnke, Hans Niels, & Michael Otte. 1981. "Origins of the Program of 'Arithmetization of Mathematics.'" In *Social History of Nineteenth Century Mathematics*, edited by Herbert Mehrtens, Henk Bos, & Ivo Schneider, 21–49. Boston, Basel, & Stuttgart: Birkhäuser.

Johnson, Ann. 2004. "From Boeing to Berkeley: Civil Engineers, the Cold War and the Origins of Finite Element Analysis." In *Growing Explanations: Historical Perspectives on Recent Science*, edited by M. Norton Wise, 133–158. Durham, NC: Duke University Press.

Johnson, Ann. 2006a. "The Shape of Molecules to Come: Algorithms, Models, and Visions of the Nanoscale." In *Simulation: Pragmatic Constructions of Reality*, edited by Johannes Lenhard, Günther Küppers, & Terry Shinn, 25–39. New York & Berlin: Springer.

Johnson, Ann. 2006b. "Infrastructure for Simulations: The Case of Computational Nanotechnology." *Science Studies* 19 (1): 34–51.

Johnson, Ann. 2009. "Modeling Molecules: Computational Nanotechnology as a Knowledge Community." *Perspectives on Science* 17 (2): 144–173.

Johnson, Ann, & Johannes Lenhard. 2011. "Towards a Culture of Prediction: Computational Modeling in the Era of Desktop Computing." In *Science*

Transformed? Debating Claims of an Epochal Break, edited by Alfred Nordmann, Hans Radder, & Gregor Schiemann, 189–199. Pittsburgh: University of Pittsburgh Press.

Jones, Caroline A., & Peter Galison, eds. 1998. *Picturing Science, Producing Art.* New York: Routledge.

Jost, Jürgen. 2017. "Object-Oriented Models vs. Data Analysis—Is This the Right Alternative?" In *Mathematics as a Tool*, edited by Johannes Lenhard & Martin Carrier, 253–286. Boston Studies in History and Philosophy of Science 327. New York: Springer.

Kac, Mark, & Stanislaw Ulam. 1968. *Mathematics and Logic.* New York: Praeger.

Kauer, Gerhard, & Helmut Blöcker. 2003. "Applying Signal Theory to the Analysis of Biomolecules." *Bioinformatics* 19: 2016–2021.

Keller, Evelyn Fox. 2003. "Models, Simulation, and 'Computer Experiments.'" In *The Philosophy of Scientific Experimentation*, edited by Hans Radder, 198–215. Pittsburgh, PA: University of Pittsburgh Press.

Kitcher, Philip, & Wesley Salmon, eds. 1989. *Scientific Explanation.* Minneapolis: University of Minnesota Press.

Knobloch, Tobias. 2009. "Quasi-Experimente. Vom Erkenntniswert von Gedankenexperiment und Computersimulation." Ph.D. dissertation, Bielefeld University.

Knuuttila, Tarja, Martina Merz, & Erika Mattila, eds. 2006. "Computer Models and Simulations in Scientific Practice." *Science Studies* 19 (1): 3–11.

Koyré, Alexandre. 1968. *Metaphysics and Measurement: Essays in the Scientific Revolution.* London: Chapman & Hall.

Krohn, Wolfgang. 2006. *Francis Bacon.* 2nd edition. München: C. H. Beck.

Krohn, Wolfgang. 2007. "Realexperimente—Die Modernisierung der 'offenen Gesellschaft' durch experimentelle Forschung." *Erwägen – Wissen – Ethik / Deliberation – Knowledge – Ethics* 18 (3): 343–356.

Krohs, Ulrich. 2008. "How Digital Computer Simulations Explain Real-World Processes." *International Studies in the Philosophy of Science* 22 (3): 277–292.

Küppers, Günther, & Johannes Lenhard. 2005. "Computersimulationen: Modellierun gen Zweiter Ordnung." *Journal for General Philosophy of Science* 36 (2): 305–329.

Küppers, Günther, Johannes Lenhard, & Terry Shinn. 2006. "Computer Simulation: Practice, Epistemology, and Social Dynamics." In *Simulation: Pragmatic Construction of Reality*, edited by Johannes Lenhard, Günther Küppers, & Terry Shinn, 3–22. Dordrecht: Springer.

Ladyman, James. 1998. "What Is Structural Realism?" *Studies in History and Philosophy of Science Part B: Studies in History and Philosophy of Modern Physics* 29: 409–424.

Landman, Uzi. 2001. "Lubricating Nanoscale Machines: Unusual Behavior of Highly Confined Fluids Challenges Conventional Expectations." *Georgia Tech Research News*, February. Online. http://gtresearchnews.gatech.edu/newsrelease/landman/landman_news.htm.

Landman, Uzi. 2005. "Studies of Nanoscale Friction and Lubrication," *Georgia Tech Research News*, October. Online. http://gtresearchnews.gatech.edu/newsrelease/MRSMEDAL.htm.

Landman, Uzi, W. D. Luedtke, Nancy A. Burnham, & Richard J. Colton. 1990. "Atomistic Mechanisms and Dynamics of Adhesion, Nanoindentation, and Fracture." *Science* 248: 454–461.

Langton, Christopher G. 1996. "Artificial Life." In *The Philosophy of Artificial Life*, edited by Margaret Boden, 39–94. Oxford: Oxford University Press.

Lehtinen, Ari, & Jaakko Kuorikoski. 2007. "Computing the Perfect Model: Why Do Economists Shun Simulation?" *Philosophy of Science* 74: 304–329.

Lenhard, Johannes. 2006. "Surprised by a Nanowire: Simulation, Control, and Understanding." *Philosophy of Science* 73 (5): 605–616.

Lenhard, Johannes. 2007. "Computer Simulation: The Cooperation Between Experimenting and Modeling." *Philosophy of Science* 74: 176–194.

Lenhard, Johannes. 2009. "The Great Deluge: Simulation Modeling and Scientific Understanding." In *Scientific Understanding. Philosophical Perspectives*, edited by Henk W. de Regt, Sabina Leonelli, & Kai Eigner, 169–186. Pittsburgh, PA: University of Pittsburgh Press.

Lenhard, Johannes. 2014. "Disciplines, Models, and Computers: The Path to Computational Quantum Chemistry." *Studies in History and Philosophy of Science Part B: Studies in History and Philosophy of Science Part A* 48: 89–96.

Lenhard, Johannes. 2015. *Mit Allem Rechnen—Zur Philosophie der Computersimulation*. Berlin: de Gruyter.

Lenhard, Johannes. 2017a. "Thought Experiments and Simulation Experiments. Exploring Hypothetical Worlds." In *The Routledge Companion to Thought Experiments*, edited by Michael Stuart, Yiftach Fehige, & James R. Brown, 484–497. New York: Routledge.

Lenhard, Johannes. 2017b. "License to Explore: How Images Work in Simulation Modeling." In *The Active Image. Architecture and Engineering in the Age of Modeling*, edited by Sabine Ammon & R. Capdevila-Werning, 233–254. Cham: Springer.

Lenhard, Johannes. 2017c. "Computation and Simulation." In *The Oxford Handbook on Interdisciplinarity*, 2nd ed., edited by Robert Frodeman, Julie Thompson Klein, & R. C. D. S. Pacheco, 443–455. New York: Oxford University Press.

Lenhard, Johannes. 2018. "Holism or the Erosion of Modularity: A Methodological Challenge for Validation." *Philosophy of Science* 85 (5).

Lenhard, Johannes, & Martin Carrier, eds. 2017. *Mathematics as a Tool*. Boston Studies in the Philosophy and History of Science 327. New York: Springer.

Lenhard, Johannes, Günther Küppers, & Terry Shinn, eds. 2006. *Simulation: Pragmatic Construction of Reality*. Sociology of the Sciences Yearbook 25. Dordrecht: Springer.

Lenhard, Johannes, & Michael Otte. 2005. "Grenzen der Mathematisierung—Von der Grundlegenden Bedeutung der Anwendung." *Philosophia Naturalis* 42 (1): 15–47.

Lenhard, Johannes, & Michael Otte. 2010. "Two Types of Mathematization." In *Philosophical Perspectives on Mathematical Practice,* edited by Bart van Kerkhove, Jonas de Vuyst, & Jean Paul van Bendegem, 301–330. London: College Publications.

Lenhard, Johannes, & Eric Winsberg. 2010. "Holism, Entrenchment, and the Future of Climate Model Pluralism." *Studies in the History and Philosophy of Modern Physics* 41: 253–262.

Leonardo da Vinci. 1958. *Philosophische Tagebücher,* edited by Guiseppe Zamboni. Hamburg: Rowohlt Verlag.

Lewis, John M. 1998. "Clarifying the Dynamics of the General Circulation: Phillips's 1956 Experiment." *Bulletin of the American Meteorological Society* 79 (1): 39–60.

Lichtenberg, Georg Christoph. 2012. *Philosophical Writings.* Edited by Stephen Tester. Albany, NY: State University of New York Press.

Lorenz, Edward. 1967. *The Nature of the Theory of the General Circulation of the Atmosphere.* Technical Paper 218. Geneva: World Meteorological Organization.

Lorenz, Edward. 1993. *The Essence of Chaos.* Seattle: University of Washington Press.

Lynch, Michael, & Steve Woolgar. 1988. *Representation in Scientific Practice.* Cambridge, MA: MIT Press.

MacKenzie, Donald. 2001. *Mechanizing Proof: Computing, Risk, and Trust.* Cambridge, MA: MIT Press.

MacLeod, Miles, & Nancy J. Nersessian. 2015. "Modeling Systems-Level Dynamics: Understanding without Mechanistic Explanation in Integrative Systems Biology." *Studies in History and Philosophy of Science Part C—Biological and Biomedical Science* 49: 1–11.

Mahoney, Michael S. 1991. "Die Anfänge der Algebraischen Denkweise im 17. Jahrhundert." *RETE: Strukturgeschichte Der Naturwissenschaften:* 15–31.

Mahoney, Michael S. 1992. "Computers and Mathematics: The Search for a Discipline of Computer Science." In *The Space of Mathematics,* edited by Javier Echeverria, Andoni Ibarra, & Thomas Mormann, 347–361. Berlin and New York: de Gruyter.

Mahoney, Michael S. 1997. "Computer Science: The Search for a Mathematical Theory." In *Companion to Science in the Twentieth Century,* edited by John Krige & Dominique Pestre, 617–634. London: Routledge.

Mahoney, Michael S. 2010. "The Structures of Computation and the Mathematical Structure of Nature." *Rutherford Journal* 3. http://www.rutherfordjournal.org/article030107.html.

Mainzer, Klaus. 2003. *Computerphilosophie. Zur Einführung.* Hamburg: Junius Verlag.

Martin, Robert C. 2009. *Clean Code: A Handbook of Agile Software Craftmanship.* Upper Saddle River, NJ: Prentice Hall.

Martz, Eric, & Eric Francoeur. 1997. "The History of Visualization of Biological Macromolecules." Website from University of Massachusetts, Amherst. http://www.umass.edu/microbio/rasmol/history.htm.

Masani, Pesi. 1990. *Norbert Wiener 1894–1964*. Basel, Boston, & Berlin: Birkhäuser.

Mauritsen, Thorsten, Bjoern Stevens, Erich Roeckner, Traute Crueger, Monika Esch, Marco Giorgetta, Helmuth Haak, et al. 2012. "Tuning the Climate of a Global Model." *Journal of Advances in Modeling Earth Systems* 4: M00A01. https://doi.org/10.1029/2012MS000154.

McCulloch, Warren, & Walter Pitts. 1943. "A Logical Calculus of the Ideas Immanent in Nervous Activity." *Bulletin of Mathematical Biophysics* 5: 115–133.

McLeod, John. 1968. *Simulation: The Dynamic Modeling of Ideas and Systems with Computers*. New York: McGraw-Hill.

Merz, Martina. 2006. "Locating the Dry Lab on the Lab Map." In *Simulation: Pragmatic Construction of Reality*, edited by Johannes Lenhard, Günther Küppers, & Terry Shinn, 155–172. Sociology of the Sciences Yearbook 25. Dordrecht: Springer.

Metropolis, Nicolas, Arianna W. Rosenbluth, Marshall N. Rosenbluth, Augusta H. Teller, & Edward Teller. 1953. "Equations of State Calculations by Fast Computing Machines." *Journal of Chemical Physics* 21 (6): 1087–1092.

Metropolis, Nicolas, & Stanislaw Ulam. 1949. "The Monte Carlo Method." *Journal of the American Statistical Association* 44: 335–341.

Mindell, David A. 2002. *Between Human and Machine: Feedback, Control, and Computing Before Cybernetics*. Baltimore & London: Johns Hopkins University Press.

Minsky, Marvin. 1967. "Why Programming Is a Good Medium for Expressing Poorly Understood and Sloppily Formulated Ideas." In *Design and Planning II*, edited by Martin Krampen & Peter Seitz, 117–121. New York: Hastings House.

Mitchell, Sandra. 2009. *Unsimple Truths: Science, Complexity, and Policy*. Chicago: University of Chicago Press.

Mody, Cyrus. 2006. "Corporations, Universities, and Instrumental Communities: Commercializing Probe Microscopy, 1981–1996." *Technology and Culture* 47: 57–80.

Morgan, Mary S. 2003. "Experiments Without Material Intervention. Model Experiments, Virtual Experiments, and Virtually Experiments." In *The Philosophy of Scientific Experimentation*, edited by Hans Radder, 216–235. Pittsburgh, PA: University of Pittsburgh Press.

Morgan, Mary S., & Margaret Morrison, eds. 1999. *Models as Mediators: Perspectives on Natural and Social Science*. Cambridge: Cambridge University Press.

Morrison, Margaret. 1999. "Models as Autonomous Agents." In *Models as Mediators*, edited by Mary S. Morgan & Margaret Morrison, 38–65. Cambridge: Cambridge University Press.

Morrison, Margaret. 2014. *Reconstructing Reality: Models, Mathematics, and Simulations*. New York: Oxford University Press.

Napoletani, Domenico, Marco Panza, & Daniele C. Struppa. 2011. "Agnostic Science. Towards a Philosophy of Data Analysis." *Foundations of Science* 16: 1–20.

Neelamkavil, Francis. 1987. *Computer Simulation and Modelling*. New York: John Wiley.

Neurosolutions. 2007. www.neurosolutions.com/products.

Nordmann, Alfred. 2006. "Collapse of Distance. Epistemic Strategies of Science and Technoscience." *Danish Yearbook of Philosophy* 41: 7–34.

Nordmann, Alfred. 2010. "Science in the Context of Technology." In *Science in the Context of Application*, edited by Martin Carrier & Alfred Nordmann, 467–482. Boston Studies in the Philosophy and History of Science 274. Dordrecht: Springer.

Oberkampf, William L., & Christopher J. Roy. 2010. *Verification and Validation in Scientific Computing*. Cambridge: Cambridge University Press.

Oberkampf, William L., & T. G. Trucano. 2000. "Validation Methodology in Computational Fluid Dynamics." In *AAIA 2000—2549*. Denver, CO: American Institute of Aeronautics and Astronautics.

O'Malley, Charles D., ed. 1968. *Leonardo's Legacy: An International Symposium*. Berkeley, CA: University of California Press.

Otte, Michael. 1994. *Das Formale, das Soziale und das Subjektive*. Frankfurt am Main: Suhrkamp.

Pahl, Gerhard, & Wolfgang Beitz. 1984/1996/2007. *Engineering Design: A Systematic Approach*. Berlin: Springer.

Parker, Wendy. 2006. "Understanding Pluralism in Climate Modeling." *Foundations of Science* 11 (4): 349–368.

Parker, Wendy. 2014a. "Simulation and Understanding in the Study of Weather and Climate." *Perspectives on Science* 22 (3): 336–356.

Parker, Wendy. 2014b. "Values and Uncertainties in Climate Prediction, Revisited." *Studies in History and Philosophy of Science Part A* 46: 24–30.

Peitgen, Heinz-Otto, & Michael F. Barnsley, eds. 1988. *The Science of Fractal Images*. New York: Springer.

Peitgen, Heinz-Otto, & Peter H. Richter. 1986. *The Beauty of Fractals: Images of Complex Dynamical Systems*. Berlin: Springer.

Perdew, John P., Adrienn Ruzsinsky, Jianmin Tao, Victor N. Staroverov, Gustavo E. Scuseria, & Gábor I. Csonka. 2005. "Prescription for the Design and Selection of Density Functional Approximations: More Constraint Satisfaction with Fewer Fits." *Journal of Chemical Physics* 123. https://doi.org/10.1063/1.1904565.

Pérez-Ramos, Antonio. 1996. "Bacon's Forms and the Maker's Knowledge Tradition." In *The Cambridge Companion to Bacon*, edited by Markku Peltonen, 99–120. Cambridge: Cambridge University Press.

Peterson, Ivars. 1995. *Fatal Defect: Chasing Killer Computer Bugs*. New York: Vintage.

Petroski, Henry. 1992. *To Engineer Is Human: The Role of Failure in Successful Design*. New York: Vintage.

Pfeffer, Richard L., ed. 1960. *Dynamics of Climate: The Proceedings of a Conference on the Application of Numerical Integration Techniques to the Problem of the General Circulation, October 26–28, 1955*. Oxford: Pergamon.

Pfeifer, Rolf, & Christian Scheier. 1999. *Understanding Intelligence*. Cambridge, MA, & London: MIT Press.

Phillips, Norman. 1956. "The General Circulation of the Atmosphere: A Numerical Experiment." *Quarterly Journal of the Royal Meteorological Society* 82 (352): 123–164.

Phillips, Norman. 1959. "An Example of Non-Linear Computational Instability." In *The Atmosphere and the Sea in Motion*, edited by Bert Bolin, 501–504. New York: Rockefeller Institute Press.

Phillips, Norman. 2000. "Foreword." In *General Circulation Model Development*, edited by David A. Randall, xxvii–xxix. San Diego, CA: Academic Press.

Price, Derek J. deSolla. 1984. "Notes Towards a Philosophy of the Science/Technology Interaction." In *The Nature of Technological Knowledge*, edited by Rachel Laudan, 105–114. Dordrecht: Reidel.

Radder, Hans, ed. 2003. *The Philosophy of Scientific Experimentation*. Pittsburgh, PA: University of Pittsburgh Press.

Radder, Hans. 2009. "Science, Technology, and the Science-Technology Relationship." In *Handbook Philosophy of Technology and Engineering Sciences*, edited by Anthonie Meijers, 67–91. Amsterdam: Elsevier.

Raible, Martin, Mykhailo Evstigneev, Frank W. Bartels, Rainer Eckel, M. Nguyen-Duong, R. Merkel, Robert Ros, Dario Anselmetti, & Peter Reimann. 2006. "Theoretical Analysis of Single-Molecule Force Spectroscopy Experiments: Heterogeneity of Chemical Bonds." *Biophysical Journal* 90: 3851–3864.

Raible, Martin, Mykhailo Evstigneev, Peter Reimann, Frank W. Bartels, & Robert Ros. 2004. "Theoretical Analysis of Dynamic Force Spectroscopy on Ligand-Receptor Complexes." *Journal of Biotechnology* 112: 13–23.

Ramsey, Jeffry L. 1997. "Between the Fundamental and the Phenomonogical: The Challenge of the Semi-Empirical Methods." *Philosophy of Science* 64: 627–653.

Ramsey, Jeffry L. 2007. "Calibrating and Constructing Models of Protein Folding." *Synthese* 155: 307–320.

Randall, John Hermann Jr. 1961. "The Place of Leonardo da Vinci in the Emergence of Modern Science." In *The School of Padua and the Emergence of Modern Science*, edited by John Hermann Jr. Randall, 115–138. Padova: Editrice Antenore.

Reifenrath, Roderich. 1999. "Die Geschichte der Simulation. Überlegungen zur Genese des Bildes am Computerbildschirm." Ph. D dissertation, TU Berlin.

Rheinberger, Hans-Jörg. 2001. "Objekt und Repräsentation." In *Mit dem Auge Denken: Strategien der Sichtbarmachung in Wissenschaftlichen und Virtuellen Welten*, edited by Bettina Heintz & Jörg Huber, 55–61. Zürich: Edition Voldemeer.

Rheinberger, Hans-Jörg. 2006. *Experimentalsysteme und Epistemische Dinge*. Frankfurt: Suhrkamp.

Richardson, Lewis F. 1922. *Weather Prediction by Numerical Process*. London: Cambridge University Press.

Rohrlich, Fritz. 1991. "Computer Simulation in the Physical Sciences." In *PSA 1990*, edited by Arthur Fine, F. Forbes, & L. Wessels, 2:507–518. East Lansing, MI: Philosophy of Science Association.

Romine, Glen. 2002. "Opal Project." National Center for Supercomputing Applications, University of Illinois, Urbana/Champaign. http://redrock.ncsa.uiuc.edu/CMG/People/romine/introd.html.

Romine, Glen, & Richard Wilhelmson. 2001. "Parameterization Impacts on a Simulation of Hurricane Opal (1995)." In Preprints, *11th PSU/NCAR Mesoscale Model User's Workshop*, Boulder, Colorado, 86–89.

Rosen, Robert. 1988. "Effective Processes and Natural Law." In *The Universal Turing Machine: A Half-Century Survey*, edited by Rolf Herken, 523–537. Hamburg: Kammerer & Unverzagt.

Rosenblueth, Arturo, & Norbert Wiener. 1945. "The Role of Models in Science." *Philosophy of Science* 12 (4): 316–322.

Royce, Wynston. 1970. "Managing the Development of Large Software Systems." *Proceedings of IEEE WESCON* 26: 1–9.

Scerri, Eric. 2004. "Just How Ab Initio Is Ab Initio Quantum Chemistry?" *Foundations of Chemistry* 6: 93–116.

Searle, John. 1980. "Minds, Brains and Programs." *Behavioral and Brain Sciences* 3: 417–457.

Seeger, Raymond J. 1974. "On the Physical Insights of Leonardo da Vinci." In *Philosophical Foundations of Science*, edited by Raymond J. Seeger & Robert S. Cohen, 41–53. Dordrecht and Boston: Reidel Publishing Company.

Seiden, Philip E., & Lawrence S. Schulman. 1990. "Percolation Model of Galactic Structure." *Advances in Physics* 39 (1): 1–54.

Shapin, Steven, & Simon Shaffer. 1985. *Leviathan and the Air-Pump. Hobbes, Boyle, and the Experimental Life*. Princeton, NJ: Princeton University Press.

Shinn, Terry. 2006. "When Is Simulation a Research Technology? Practices, Markets, and Lingua Franca." In *Simulation: Pragmatic Construction of Reality*, edited by Johannes Lenhard, Günther Küppers, & Terry Shinn, 187–203. Dordrecht: Springer.

Simon, Herbert A. 1969. *The Sciences of the Artificial*. Cambridge, MA: MIT Press.

Sismondo, Sergio, & Snait Gissis, eds. 1999. "Practices of Modeling and Simulation." *Science in Context* 12 (Special Issue): 247–376.

Smith, Leonard A. 2002. "What Might We Learn From Climate Forecasts?" *Proceeedings of the National Academy of Sciences USA* 4 (99): 2487–2492.

Solomon, S., D. Qin, M. Manning, Z. Chen, M. Marquis, K. B. Averyt, M. Tignor, & H. L. Miller, eds. 2007. *Contribution of Working Group I to the Fourth Assessment Report of the Intergovernmental Panel on Climate Change*. Cambridge & New York: Cambridge University Press.

Stainforth, David A., Thomas E. Downing, Richard Washington, & Mark New. 2007. "Issues in the Interpretation of Climate Model Ensembles to Inform Decisions." *Philosophical Transactions of the Royal Society A* 365 (1857): 2145–2161.

Steels, Luc, & Rodney Brooks, eds. 1995. *The Artificial Life Route to Artificial Intelligence: Building Embodied, Situated Agents*. Hillsdale, NY: Lawrence Erlbaum.

Steinle, Friedrich. 2002. "Discovering? Justifying? Experiments in History and Philosophy of Science." In *Revisiting Discovery and Justification*, edited by Friedrich Steinle & Jutta Schickore, 175–184. Preprint 211. Berlin: Max-Planck-Institut für Wissenschaftsgeschichte.

Steinle, Friedrich. 2005. *Explorative Experimente. Ampère, Faraday und die Ursprünge der Elektrodynamik*. Stuttgart: Steiner.

Stevens, Bjoern, & Sandrine Bony. 2013. "What Are Climate Models Missing?" *Science* 340: 1053–1054.

Stöckler, Manfred. 2000. "On Modeling and Simulations as Instruments for the Study of Complex Systems." In *Science at Century's End*, edited by Martin Carrier, Gerald J. Massey, & Laura Ruetsche, 355–373. Pittsburgh, PA: University of Pittsburgh Press.

Suárez, Mauricio. 2004. "An Inferential Conception of Scientific Representation." *Philosophy of Science* 71: 767–779.

Sundberg, Mikaela. 2007. "Parameterizations as Boundary Objects on the Climate Arena." *Social Studies of Science* 37 (3): 473–488.

Suppes, Patrick. 1961. "A Comparison of the Meaning and Use of Models in the Mathematical and Empirical Sciences." In *The Concept and Role of the Model in Mathematics and Natural and Social Sciences*, edited by Hans Freudenthal, 163–177. Dordrecht: Reidel.

Toulmin, Stephen. 1990. *Cosmopolis. The Hidden Agenda of Modernity*. Chicago: The University of Chicago Press.

Turing, Alan M. 1950. "Computing Machinery and Intelligence." *Mind* LIX (236): 433–460.

Turing, Alan M. 1986. "Proposed Electronic Calculator. Report for National Physical Laboratory, Teddington". In *A. M. Turing's A. C. E. Report of 1946 and Other Papers*, edited by Brian E. Carpenter, & Robert W. Doran. Cambridge, MA: The MIT Press.

Turkle, Sherry. 2009. *Simulation and Its Discontents*. Cambridge, MA, & London, England: The MIT Press.

Tymoczko, Thomas. 1979. "The Four-Color Problem and Its Philosophical Significance." *Journal of Philosophy* 76 (2): 57–83.

Ulam, Stanislaw. 1952. "Random Processes and Transformations." In *Proceedings of the International Congress of Mathematicians 1950*, 264–275. Providence: American Mathematical Society.

Ulam, Stanislaw. 1986. *Science, Computers, and People: From the Tree of Mathematics*. Boston, Basel, & Stuttgart: Birkhäuser.

Urbach, Peter. 1987. *Francis Bacon's Philosophy of Science. An Account and a Reappraisal*. La Salle, IL: Open Court.

van Kerkhove, Bart, & Jean Paul van Bendegem. 2008. "Pi on Earth, or Mathematics in the Real World." *Erkenntnis* 68 (3): 421–435.

van Fraassen, Bas C. 1980. *The Scientific Image*. Oxford: Oxford University Press.

Vincenti, Walter G. 1990. *What Engineers Know and How They Know It*. Baltimore, MD: Johns Hopkins University Press.

von Neumann, John. 1951. "The General and Logical Theory of Automata." In *Cerebral Mechanisms in Behavior—The Hixon Symposium*, edited by L. A. Jeffress, 1–31. New York: John Wiley.

von Neumann, John. [1945]1993. "First Draft of a Report on the EDVAC." IEEE Annals of the History of Computing 15 (4): 27–75.

von Neumann, John, & Robert D. Richtmyer. 1947. "Statistical Methods in Neutron Diffusion." In *Analogies Between Analogies: The Mathematical Reports of S. M. Ulam and His Los Alamos Collaborators*, edited by Stanislaw Ulam, A. R. Bednarek, & Francoise Ulam, 17–36. Berkeley & Los Angeles: University of California Press.

Weart, Spencer, Margaret Jacob, & Harold J. Cook. 2003. *The Discovery of Global Warming*. Cambridge, MA: Harvard University Press.

Weingart, Peter, Martin Carrier, & Wolfgang Krohn, eds. 2007. *Nachrichten aus der Wissensgesellschaft*. Weilerswist: Velbrück.

Weisberg, Michael. 2013. *Simulation and Similarity: Using Models to Understand the World*. New York: Oxford University Press.

Weizenbaum, Joseph. 1976. *Computer Power and Human Reason: From Judgment to Calculation*. San Francisco: W. H. Freeman.

Wiener, Norbert. 1948. *Cybernetics or Communication and Control and the Animal and the Machine*. Cambridge, MA: MIT Press.

Wiener, Norbert. 1954. "Nonlinear Prediction and Dynamics." In *Proceedings of the Third Berkeley Symposium on Mathematical Statistics and Probability* 3: 247–252.

Wiener, Norbert. 1956. *I Am a Mathematician*. Cambridge, MA, & London: MIT Press.

Wiener, Norbert. 1960. "Some Moral and Technical Consequences of Automation." *Science* 131: 1355.

Wiin-Nielsen, Aksel. 1991. "The Birth of Numerical Weather Prediction." *Tellus* 43: 36–52.

Wilholt, Torsten. 2006. "Design Rules: Industrial Research and Epistemic Merit." *Philosophy of Science* 73 (1): 66–89.

Wimsatt, William C. 2007. *Re-Engineering Philosophy for Limited Beings: Piecewise Approximations to Reality*. Cambridge, MA, & London: Harvard University Press.

Winograd, Terry, & Fernando Flores. 1991. *Understanding Computers and Cognition*. 5th ed. Reading, MA: Addison-Wesley.

Winsberg, Eric. 1999. "Sanctioning Models: The Epistemology of Simulation." *Science in Context* 12 (2): 275–292.

Winsberg, Eric. 2003. "Winsberg, Eric, Simulated Experiments: Methodology for a Virtual World." *Philosophy of Science* 70: 105–125.

Winsberg, Eric. 2006a. "Models of Success versus the Success of Models: Reliability Without Truth." *Synthese* 152: 1–19.

Winsberg, Eric. 2006b. "Handshaking Your Way to the Top: Inconsistency and Falsification in Intertheoretic Reduction." *Philosophy of Science* 73: 582–594.

Winsberg, Eric. 2010. *Science in the Age of Computer Simulation*. Chicago: University of Chicago Press.

Wise, M. Norton, ed. 2004. *Growing Explanations. Historical Perspectives on Recent Science*. Durham, NC, & London: Duke University Press.

Wise, M. Norton. 2007. "Science *Is* Technology? Comment on Paul Forman, given at SHOT Meeting, 20. Oktober 2007." 9 page manuscript, personal communication.

Wolfram, Stephen. 2002. *A New Kind of Science*. Champaign, IL: Wolfram Media.

Worrall, John. 1989. "Structural Realism: The Best of Both Worlds?" *Dialectica* 43: 99–124.

Ylikoski, Petri. 2014. "Agent-Based Simulation and Sociological Understanding." *Perspectives on Science* 22 (3): 318–335.

Zeigler, Bernard P. 1976. *Theory of Modelling and Simulation*. Malabar: Krieger.

INDEX

www.ingramcontent.com/pod-product-compliance
Lightning Source LLC
Chambersburg PA
CBHW051213090826
49497CB00083B/114
9780190873288